Frequently Used Formulas

Chapter 2

1. $m = \dfrac{\Sigma X}{n}$

2. $s = \sqrt{\dfrac{\Sigma(X - m)^2}{n - 1}}$

3. $s = \sqrt{\dfrac{\Sigma X^2 - \dfrac{(\Sigma X)^2}{n}}{n - 1}}$

4. $z_X = \dfrac{X - \mu}{\sigma}$, $z_X = \dfrac{X - \mu}{s}$, $z_X = \dfrac{X - m}{s}$

5. $X = \mu + z\sigma$, $X = \mu + zs$, $X = m + zs$

6. $PR_X = \dfrac{B + \frac{1}{2}E}{n}$ (100)

Chapter 4

1. $p = P(\text{event}) = \dfrac{F}{T}$

2. $q = 1 - p$

Chapter 7

1. $\mu_s = np$

2. $\sigma_s = \sqrt{npq}$

3. $S_? = \mu_s + z\sigma_s$

Chapter 8

1. $\mu_s = np$

2. $\sigma_s = \sqrt{npq}$

3. $S_c = \mu_s + z_c\sigma_s$

4. Power $= 1 - \beta$

Chapter 9

1. $\mu_{d\hat{p}} = p_1 - p_2$

2. $\hat{\sigma}_{d\hat{p}} = \sqrt{\dfrac{\hat{p}\hat{q}}{n_1} + \dfrac{\hat{p}\hat{q}}{n_2}}$

3. The sample difference $d\hat{p} = \hat{p}_1 - \hat{p}_2$

4. The critical difference $d\hat{p}_c = \mu_{d\hat{p}} + z_c\hat{\sigma}_{d\hat{p}}$

Chapter 10

ONE-SAMPLE TESTS

1. $\mu_m = \mu_{\text{pop}}$

2. $s = \sqrt{\dfrac{\Sigma X^2 - \dfrac{(\Sigma X)^2}{n}}{n - 1}}$

(Continued on back cover)

UNDERSTANDING STATISTICS

SECOND EDITION

UNDERSTANDING STATISTICS

ARNOLD NAIMAN

ROBERT ROSENFELD
Nassau Community College

GENE ZIRKEL
Nassau Community College

McGRAW-HILL BOOK COMPANY

New York   St. Louis   San Francisco   Auckland
Bogotá   Düsseldorf   Johannesburg   London   Madrid
Mexico   Montreal   New Delhi   Panama   Paris
São Paulo   Singapore   Sydney   Tokyo   Toronto

67890 DODO 8321

This book was set in Helvetica by Progressive Typographers.
The editors were A. Anthony Arthur,
Alice Macnow, and Shelly Levine Langman;
the designer was Nicholas Krenitsky;
the production supervisor was Leroy A. Young.
New drawings were done by B. Handelman Associates, Inc.
R. R. Donnelley & Sons Company was printer and binder.

Library of Congress Cataloging in Publication Data
Naiman, Arnold.
    Understanding statistics.

    Includes index.
    1. Statistics. I. Rosenfeld, Robert, joint author.
II. Zirkel, Gene, joint author. III. Title.
HA29.N28 1977   519.5   76-20574
ISBN 0-07-045860-x

In memory of our colleague
Dr. ARNOLD NAIMAN

Contents

Preface
to the Second Edition

Why a second edition? The answer, quite simply, is that our colleagues have suggested many improvements; teachers and students using the book have suggested changes and additions which we would like to make. We thank them all. We are indebted to our friends at Nassau Community College, at McGraw-Hill, and to the many other users of this book who took the time to comment. But we must mention specially the contribution of our coauthor, the late Dr. Arnold Naiman, and his wife, Inez.

Arnie died while we were working on this revision. His many notes, corrections, and revisions are an integral part of this edition, and we are grateful to Inez for giving them to us. We feel that Arnie would have been pleased with the way this revision turned out, and so we have dedicated it to his memory.

Past users of the book will find most of the old material and all of the old light approach still here. However, a few examples have been changed in light of new ideas of sexual stereotyping. (It was sad to see Pamela Passion go.)

We have rewritten Chapter 8, the key chapter introducing hypothesis testing. We have added new material to the chapter on chi square and the chapter on confidence intervals. We changed the chapter on two-sample binomial tests so that the problems can be done by the "pooled-sample" method.

We have added two chapters. Chapter 15 covers tests involving variances, including a very light introduction to analysis of variance. Chapter 16 is about some simple nonparametric tests.

At the request of many teachers more exercises have been added, as well as some introductory field projects in the early chapters. Some of the new exercises are of a slightly different style, and a number are more difficult.

Some small changes were suggested only once, but they were good and so we made them. We now use the combination symbol ($\binom{8}{3}$) when referring to Pascal's Triangle. Also, we have moved the list of formulas from an appendix to the more convenient end papers.

All in all we hope that these changes keep the good points of the first edition intact while satisfying the requests of those who suggested them. We feel that the book has been improved. We hope that you will continue to inform us of your views.

The difficult and often frustrating job of typing our manuscript was done by Carmen DeCosta, who not only did an outstanding job technically, but who several times found and corrected errors we had made. Special thanks must go to her.

For helpful, detailed criticism of the manuscript we would like to thank Professors James Edmondson, Santa Barbara City College; John S. Mowbray, Shippensburg State College; and Paul Kroll, formerly of William Paterson College of New Jersey. Their suggestions helped us to clarify and better organize the text.

Robert Rosenfeld
Gene Zirkel

Preface
to the First Edition

Having taught statistics for several years (principally at Nassau Community College), we have worked with students who have very little mathematics background. Therefore, we decided to write a book that would be elementary enough to reach this audience and still be mathematically sound. Our intention was to cover this material in a one-semester college level course. With open enrollment coming to many colleges and the admission of many students without the usual academic background, the need for such a text is increasing. "Understanding Statistics" is designed to fill this need.

The objective of this book is to show the readers *how* statistics is used, not to train them to be statisticians. Students will gain an appreciation of the proper use of statistics and statistical terms that confront them in textbooks, newspapers, magazines, and on TV and radio. The major emphasis of the book is on understanding sampling and hypothesis testing.

We feel that the best way to introduce inferential statistics is through probability theory. Therefore, after a brief discussion of descriptive statistics, probability is treated intuitively. Our treatment of probability leads into the binomial distribution and then the normal distribution is initially introduced as an approximation to the binomial distribution. This brings us to Chapter 8. Here we bring together all the material presented thus far and discuss the method of statistical hypothesis testing. One-sample binomial tests are used to introduce this important idea.

The next few chapters discuss other types of hypothesis testing. Two-sample binomial, one- and two-sample tests of sample means with both large samples (*z* scores) and small samples (*t* scores), and chi-square tests are included. A chapter on correlation and prediction concludes the text. Once the students have mastered the basic material in Chapter 8, the instructor can select from the remaining chapters those topics which are appropriate to the needs of his or her students.

From Chapter 8 onward field projects are suggested after each set of exercises. We have found that students who perform hypotheses tests of their own get a much better grasp of the subject. This was found to be particularly true of the two-sample binomial tests discussed in Chapter 9.

Numerous examples and exercises are provided from various fields, such as biology, medicine, psychology, education, and political science. They have been deliberately chosen to arouse student interest. They range from the frivolous to the serious, and are not just lists of numerical exercises. The exercises are plentiful and the answers to odd-numbered exercises are given at the end of the book. A glossary of new words, symbols, and formulas is given at the end of each chapter.

Because it is advisable that the student know what mathematical skills will be needed for the material in this book, an appendix containing a selection of typical arithmetic problems is included. We strongly recommend that each student do these problems at the beginning of the course. The student should be able to handle signed numbers, but no manipulative skills from algebra are needed.

No formal proofs are presented in the text. When feasible, theorems are motivated by an appeal to common sense. While this presentation is not mathematically rigorous, care is taken that the material is at all times mathematically accurate. Topics are introduced informally by questions and examples which lead naturally to development of the pertinent ideas. Notation is kept as simple as possible, and illustrations are used throughout for clarification.

The book is easy to read and has been well received by students in preliminary editions for several semesters. The tables have been designed for ease of use, not for brevity.

We wish to thank our colleagues at Nassau Community College for their encouragement and helpful suggestions, particularly Professors Abraham Weinstein, Frank Avenoso, James Baldwin, Eli Berlinger, Alice Berridge, Mauro Cassano, Dennis Christy, Jerry Kornbluth, Roy McLeod, George Miller, and Michael Steuer. We would also like to thank Professors Daniel Brunk and Wilfrid Dixon for their critical reviews of the manuscript.

Arnold Naiman
Robert Rosenfeld
Gene Zirkel

Introduction

1

You are a statistic. Certainly you must have had the feeling at some time that you were being treated merely as a collection of numerical information.

Social security number: 198-31-1755

ZIP code: 11530

Area code: 516

Height: 5 feet 7 inches

Weight: 162 pounds

Student identification number: 4321058

You should understand how statistics are used because decisions that affect you personally are based on statistics. "Your grade-point average is only 1.47; sorry kid, we'll have to put you on probation. I know that you just had a bad time with your parents, but let's face it, everybody's got problems." That hurts.

There are probably times when you have mixed feelings about being treated as a source of statistics.

DEAR FRIEND:
    ØUR RELIABLE CØMPUTERIZED MATCHING SYSTEM HAS PRØC-
ESSED YØUR VITAL STATISTICS AND WE ARE PLEASED TØ EN-
CLØSE HEREWITH THE NAMES, ADDRESSES (WITH ZIP CØDES)
AND PHØNE NUMBERS (WITH AREA CØDES) ØF SIX IDEAL

MATCHES. WE ARE HØPEFUL THAT YØU CAN ESTABLISH A LASTING RELATIØNSHIP WITH AT LEAST ØNE ØF THEM.
WE REMIND YØU THAT YØUR FEE IS NØT REFUNDABLE.
SINCERELY,
CØMPUTERIZED DATING SERVICE

Products are sold to you all the time with numbers thrown at you: (1) "I used Grit toothpaste, and now I have 20 percent fewer cavities." (Fewer than what?) (2) "Hey, kids! Start your day with Daystart Cereal! It has twice as much iron as a delicious slice of toast and more vitamin C than two slices of bacon!" (So who said that bread was a good source of iron in the first place, or that bacon has a lot of vitamin C?)

Doctors prescribe medicine and treatment for you, basing their judgment on statistical information. (1) Use of this pill will cause deleterious side effects in 1.4 percent of its users. (Is the risk worth taking?) (2) There is a 40 percent chance that an adult suffering from a herniated spinal disk will recover spontaneously. (Should we go ahead with the operation?)

Consider the following example:

A random sample of adults was taken in a large shopping plaza in the city of Albany. Of those questioned, 15 percent used NoCav brand toothpaste. Subsequently a concentrated advertising campaign was undertaken to sell NoCav to the public. A second survey taken 3 weeks after this campaign showed that 19 percent of those questioned used NoCav toothpaste.

Are we correct in assuming that the rise from 15 percent in the first sample to 19 percent in the second sample is due to advertising? If we have doubts that the advertising caused substantial increase in the use of NoCav toothpaste, what questions should we ask concerning the data presented? What about the data that were omitted from the presentation?

TWO USES OF STATISTICS

In the NoCav example above we see numbers used in two different ways. The number 15 percent is used to describe the fraction of people in the first sample who used NoCav. As such it describes with conciseness and clarity the unreported fact that of 140 persons interviewed, 21 used NoCav. This is an example of descriptive statistics. **Descriptive statistics** is the use of numbers to summarize information which is known about some situation. In contrast to this use of numbers, if we use this sample to imply that approximately 15 percent of *all* the adults in Albany used NoCav, then we are using the number to infer something from this sample about a larger population for which we do not have complete information. This is an example of statistical inference. **Statistical inference** is the use of numbers to give numerical information about larger groups than those from which the original raw data were taken.

In characterizing a large amount of data by a few descriptive statistics we gain clarity and compactness, but we lose detail. The following statistics, by summarizing information, describe in some way the populations from which they were taken.

1. The average IQ at Nostrum College is 109.

2. The marks on the last exam ranged from 51 to 98.

3. Nielsen reports that 25 percent of those who were interviewed watched the President's news conference last Sunday night.

The following are examples of statistical inference. We might infer from appropriate samples that

1. Between 20 and 25 percent of American college students are married.

2. Cholesterol level and heart trouble are related.

3. 25 percent of all television viewers watched the President's news conference last Sunday.

Here is a more detailed example of statistical inference:

Suppose there were a disease in which three-fourths of the patients recovered without treatment within 3 months of contracting the disease. Suppose also that a doctor claims to have discovered a new drug to cure this disease. We shall administer the drug to 100 patients. Even if the drug were useless we would still expect about 75 (three-fourths) of these people to recover. Due to chance variations more or less than 75 may recover.

One of the problems of statistical inference for the example given above is to decide how many must recover before we are willing to accept the drug as a cure. Certainly if all 100 recovered we would be enthusiastic about the drug's potential. But how about 95 or 90 or 80 recoveries? Where should we draw the line?

The job of deciding where to draw the line is an important one for the statistician. It is one of the main skills we hope you will develop from this book. Did the new drug save lives or did this result occur by chance? Even if all 100 recovered, it is possible (though very unlikely) that they would have recovered anyway just by the chance selection of our sample.

It is important for the statistician to pick his or her sample in an impartial way. If by chance we happened to test the drug on only mild cases, our results would be misleading. We would hope that the sample is truly a mirror of the population we want to learn about (in this case all victims of this disease).

Sample surveys, polls, and statistical tests have become a part of our way of life. Every day, people present figures to prove or disprove some claim: Do cyclamates cause cancer? Does marijuana smoking lead to heroin use? In this book we will study some of the tests that statisticians use when making claims. We hope to show you how such tests should properly be done and how to interpret the "proof" of such claims.

SOME STATISTICAL TERMS

If we are listing the ages of students in a certain school, then each age is called a **raw score.** In general, a raw score is any number as it originally appears in an experiment. A collection of such scores is frequently referred to as a **distribution** of scores. If we consider the grades that your class gets on the first test in this course, then you will be very interested not only in the entire distribution of the class's grades, but also in one particular raw score, namely, your own grade on the test.

A scientist is trying to determine the average weight of 1-year-old male white rabbits which are raised in laboratories. It is manifestly impossible for her to weigh every rabbit in the population. If she selects 50 rabbits and determines their average weight, these 50 would be referred to as a sample from the population.

This word **population** refers to all the persons, objects, scores, or measurements under consideration. The word **sample** refers to any portion of the population. A population may be large or small.

We used the word *random* in examples at the beginning of this chapter. No word is more important to the theory of inferential statistics than this word. An item is chosen "at random" from a population if in the selection every item in the population has the same probability or chance of being selected; the process of selection does not favor any particular item either intentionally or inadvertently. A sample of items in which each item is chosen this way is called a **random sample.** Complete textbooks have been written describing procedures for selecting random samples, and the process can become quite technical. In this book we prefer to leave the idea to your intuition. It will be sufficient to think of a random sample as one which has been picked "fairly"—without prejudicing the chances of any member of the population to be chosen. For example, if we want to pick a random sample of 20 people from some population, then every possible grouping of 20 people should have an equal chance of being selected as the sample. The practice of putting paper slips into a large drum, mixing them well, and then picking one without looking is a simple model of random selection.

Statistical testing is frequently based upon the assumption that the sample was picked randomly. If it turns out that the sample was not random, the results may not be useful. Hidden, unsuspected bias can completely destroy the usefulness of statistical information and statistical inferences made from such information. For example, if random phone calls were made at 1 P.M. to sample the population of all voters, many people with full-time day jobs would be missed.

RANDOM NUMBER TABLES

We will describe one way to pick a random sample which you may find useful if you do some experiments of your own. This method uses what is called a **random number table.** The table is very useful for picking samples from small populations. For example, suppose your class has 30 students and you want to split them randomly into two groups of 15. You can use a random number table to decide which students go into group I and which go into group II.

The random number table printed in Table B-15 consists of a sequence of digits which are in an apparently random order. The table was generated by an electronic computer using instructions designed to produce a list which passes a series of statistical tests. A list of numbers which passes all these tests looks like a string of random numbers even though the computer can be made to produce the exact same list again. For example, some of the tests require that in the entire list there be about the same number of 0s as 1s, as 2s, etc.; that the number immediately after a 1 is just as likely to be a 0 as a 1, as a 2, as a 3, etc. These lists of numbers are sometimes called "pseudorandom" (imitation random) because they can be reproduced by someone who understands how the computer instructions are designed, but for all practical purposes the table can be treated as a genuinely random list. To use the table start anywhere in it by putting your finger somewhere on the page and then just read to the right. For example, suppose you come to this sequence:

42167   93093   06243   61680   07856   16376

How can you use it to decide who of the 30 students in your class goes in group I and who goes in group II?

First write down each student's name.

names	A B C D E F G H I J K L M N O P Q R S . . .
numbers	

Then read the random number string looking for numbers from 1 to 30. Since 30, the biggest number, is a 2-digit number, read the string as if it were a string of 2-digit numbers: 42 16 79 30 93 06 24 36 · · · . Then pick any of those values which are from 1 to 30, and fill them in on the name chart, continuing until the chart is complete.

names	A B C D E F G H I J K L M N O P Q R S . . .
numbers	16 30 6 24

Then you could put the students with numbers 1 to 15 in group I and those with numbers 16 to 30 in group II.

Another way to find a starting point in the table is to put your finger on the page and take the first acceptable 2-digit number as a row indicator and the second acceptable 2-digit number as a column indicator. For example, your finger is at the string 47 50 08 21 43 60 04 01 · · · . Since there are 50 rows and 10 columns in the table, the row indicator is 47 and the column indicator is 8. Therefore, you should start your search for random numbers with the number that is in the 47th row in the 8th column.

EXAMPLE 1-1   Use the following string of random numbers to split this list of 6 people into 3 random samples of size 2.

Random numbers   96200   80566   24068   57114   33895

person	number
Tude, A.	
Nein, B.	
Lecht, C.	
Nye, D.	
Vicht, E.	
Ort, F.	

SOLUTION   We will assign these people numbers from 1 to 6. Then numbers 1 and 2 will be in the first sample; 3 and 4 will be in the second sample; 5 and 6 will be in the third sample. Since the largest value, 6, is a 1-digit number, we will read the list of random numbers as a string of 1-digit numbers, and will continue until we have found the numbers 1 through 6. We get: 6, 2, 5, 4, 1, 3. Therefore,

person	number
Tude, A.	6
Nein, B.	2
Lecht, C.	5
Nye, D.	4
Vicht, E.	1
Ort, F.	3

We end up with

sample 1	sample 2	sample 3
Vicht	Ort	Lecht
Nein	Nye	Tude

GLOSSARY

VOCABULARY

1. Descriptive statistics   2. Statistical inference   3. Raw score
4. Distribution   5. Population   6. Sample
7. Random   8. Random sample

EXERCISES

1-1   95 percent of the people who use heroin started out using marijuana regularly. Therefore, using marijuana regularly leads to using heroin. Comment.

1-2   98 percent of the people who use marijuana first drank milk on a regular basis. Therefore, drinking milk on a regular basis leads to using marijuana. Comment.

1-3   Classify each of the following as either statistical inference or descriptive statistics.

(a) Walter Krankrite predicts the results of an election after looking at the votes in 15 of 100 districts.

(b) Dr. Bea Kareful, an ecologist, says that the flesh of fish in a certain lake contains an average of 400 units of mercury.

(c) At Webelo Normal High School last year the average SAT score was 528.

(d) The safety councils of Pessam and Mystic Counties predict 600 automobile accidents for the next July fourth weekend.

(e) Last year 72 percent of the workers in Scrooge and Marley's accounting firm missed at least 1 day of work.

1-4   For each of the following statements describe the *population* or *populations* that should have been sampled to get this information.

(a) 60 percent of all suicides are widows.

(b) Malignant tumors were found in 80 percent of the rats injected with 10 ml of chemical X.

(c) English majors at Hudson University have higher grade-point averages than chemistry majors.

(d) Too much cholesterol is bad for your heart.

(e) IQ scores are determined by environment.

1-5 Have someone in the class secretly mix in a large bag any amount of dried yellow split peas with any amount of dried green split peas. Without counting or even seeing *all* the peas, discuss any methods that could be used to estimate what fraction of the peas in the bag are green.

1-6 Find some uses of statistics (sample, average, percentile, etc.) in texts that you use in other courses. Can you classify them as either descriptive or inferential?

1-7 Find some uses of statistics in current magazines and newspapers. Can you classify them as descriptive or inferential?

1-8 An advertisement states that three-fourths of doctors interviewed recommended Brand *X*. What is your reaction?

1-9(a) Use the random number table to get a random arrangement of the numbers from 0 to 9.

(b) How many people in your class got the same arrangement? (There are more than 3 million possible arrangements.)

1-10 Devise a method (different from the one in the text) to split a group of 30 students into 2 equal-sized groups using a random number table.

1-11 Professor Gotcha determines grades randomly by assigning each student on his roster a 2-digit random number and then giving grades so that about 10 percent get A, 20 percent get B, 40 percent get C, 20 percent get D, and 10 percent fail. Find out how many students are in your class and where you are on the class roster. Apply Professor Gotcha's method and determine your grade. *Hint:* One way is for students with numbers from 00 to 09 to get A, 10 to 29 to get B, 30 to 69 to get C, 70 to 89 to get D, 90 to 99 to fail.

1-12 Most random number tables are constructed by a computer, but you could make one by placing 10 pieces of paper numbered 0, 1, 2, . . . , 10 into a container and selecting one at a time randomly, each time replacing the number picked before selecting the next one.

(a) Devise another method for constructing a random number table.

(b) Construct a random number table with 20 digits. Are half of them odd?

1-13(a) How could you use the random number table (Table B-15) to find some 3-digit numbers which are less than 600?

(b) Use Table B-15 to find four such numbers.

1-14(a) How could you use a random number table to simulate (imitate) 50 tosses of a fair coin?

(b) Perform part (a) above. Did you get about 25 heads?

(c) How could you use a random number table to simulate 10 tosses of a fair die? Do it.

(d) How could you use a random number table to simulate 20 tosses of a *biased* coin, if the coin were to come up heads about 80 percent of the time? Do it.

(e) A spinner is divided into 7 equal areas colored indigo, fuchsia, indigo, ochre, indigo, mauve, and ochre. How could you simulate 14 spins of this spinner with a random number table? Do it.

(f) How could you use a random number table to simulate 10 tosses of a *pair* of fair dice? Do it.

1-15   Answer (a), (b), or (c).

(a) Find some reference to the 1936 survey by *Literary Digest* which predicted that Alf Landon would easily win the United States presidential election (e.g., Huff, *How to Lie With Statistics*).

(b) Find some reference to the polls on the June 18, 1970, British election. (Check newspapers of that week.)

(c) Find some references which discuss the randomness of the December 1, 1969 draft lottery. (Check newspapers of that week.)

FIELD PROJECT

1(a) Suppose you had a random sample of students at this school. If the samples were truly random by age and by sex, then we would be confident that the average age of our sample would be close to the average age of the entire population and that the proportion of males in our sample would be near the proportion of males in the population. Your assignment, if you decide to accept it, is to devise a method of obtaining a sample of 100 students so that the age and sex will be random. Outline this method in a clear, detailed, and specific paragraph. Include the exact questions you will ask. Comment on some of the strengths and some of the weaknesses of your method.

(b) After your method has been approved by your instructor, gather these data. Include in your report the data, the average age of your sample, the number of males and the number of females in your sample, and comment on anything that occurred that was not expected. Do you think that the average age of all the students is close to the average you computed for your sample? Do you think that the proportion of males on campus is close to the proportion that you picked?

Common Statistical Measures

2

MEASURES OF CENTRAL TENDENCY

Tommy Tufluque just got his first D. He complained to the head of the mathematics department that Professor Noays grades too low. The grades on the first test were as follows:

100   100   100   63   62   60   12   12   6   2   0

Tommy indicated that the class average was 47, which he felt was rather low. Professor Noays stated that nevertheless there were more 100s than any other grade. The department head said that the middle grade was 60, which was not unusual.

Each of these three people was looking for one number to represent the general trend of these test grades. Such a number is called an average or a **measure of central tendency.** Mr. Tufluque used the **mean** or arithmetic average, which is obtained by adding the grades and dividing by the number of grades. Professor Noays used the **mode,** which is the most frequent number. The department head used the **median,** which is the middle number when the group of numbers is written in numerical order.

These are three commonly used averages. Which of them is the best? That depends on the particular situation. Consider these nine numbers: 71, 71, 71, 71, 73, 74, 74, 75, and 95. If they represent style numbers of dresses sold today in the Chic Dress Boutique, you can see that the style number 71 was the most popular. It was the mode. This would be important in reordering stock. If they represent grades from a psychology final exam, then perhaps you would want the mean, 75, for use in certain statistical testing. If they represent the annual salary, in hundreds of dollars, for the employees of Smith's Emporium, then you might take the median,

$7,300, as the average salary. Note that the mean salary of $7,500 is larger than seven of the nine salaries.

Each average has certain properties. Depending on the context, these properties may or may not be useful. For example, the median is less affected by extremely large or extremely small values, while the mean is affected by every score. In this book we will generally use the mean because it lends itself to statistical testing.

In the example above we found the median of a distribution with an *odd* number of raw scores. Finding the number in the middle was no problem. For example, the median of 3, 7, 5, 6, 8 is _____? Hopefully, you did not say 5 since the definition of median indicates the middle term *when the numbers are arranged according to size.* Thus the median of 3, 5, 6, 7, 8 is 6. If the distribution includes an *even* number of raw scores, then there are two scores in the middle and the median is defined as their arithmetic mean. Example: The median of 3, 3, 5, 6, 8, 13 is found by adding the two middle numbers, 5 and 6, together and dividing by 2. Thus, the median is 5.5. Note that half the scores are less than 5.5 and the other half are greater.

SYMBOLS AND FORMULAS

We will use n to indicate the number of numbers or raw scores in a given distribution. For the distribution 3, 1, 8, 9, $n = 4$.

For the mean of a sample we use the letter m. For the mean of the population we use the Greek letter for m, which is μ (read: mu). Measures of a population, that is, measures that take into account *every* member of the population, are called **parameters;** μ is a parameter. Measures based on sample data, that is, measures which take into account only *some* members of the population, are called **statistics;** m is a statistic.

Generally, in this book we shall use Greek letters for parameters and English letters for statistics. Thus, μ is used for the population mean, and m is used for the sample mean. It is important for you to learn how to write the Greek letters and to know their names.

We will use the capital letter X to stand for the list of numbers in a distribution. For example, X: 1, 5, 3, 2. (At times we will use X to name just one of these numbers, such as $X = 3$. Some texts use X_i for this purpose.) It will be clear from the context which way X is being used. We will use the Greek letter for capital S, Σ (it looks like a sideways M, and is read: sigma), to stand for the command "sum." Thus, if a distribution labeled X consists of the four numbers 1, 5, 3, and 2, then ΣX is $1 + 5 + 3 + 2 = 11$. The mean of this distribution is 2.75. A formula for the mean is

$$m = \frac{\Sigma X}{n} = \frac{1 + 5 + 3 + 2}{4} = \frac{11}{4} = 2.75$$

A second way of interpreting the symbols ΣX is if in a column labeled X we have the numbers 1, 5, 3, and 2, then ΣX means the sum of numbers in this column. Similarly, X^2 will indicate a column of numbers obtained by squaring each number in column X. The column X^2 will consist of 1, 25, 9, and 4; and $\Sigma X^2 = 39$. The symbol $(\Sigma X)^2$ represents $(11)^2$ or 121. Notice that $(\Sigma X)^2$ is different from ΣX^2.

X	X^2
1	1
5	25
3	9
2	4
$\Sigma X = 11$	$\Sigma X^2 = 39$

In the same way $X - 1$ will be the heading for the column obtained by subtracting 1 from each number in column X. The column will consist of 0, 4, 2, and 1. Therefore, $\Sigma(X - 1) = 7$.

X	$X - 1$
1	0
5	4
3	2
2	1
$\Sigma X = 11$	$\Sigma(X - 1) = 7$

Sometimes we want to differentiate between two different populations in the same example. If we label the first population X and the second Y, then the number of elements in the first population would be n_X (read: n sub X) and in the second population n_Y (read: n sub Y). The mean of the first would be μ_X and the mean of the second μ_Y. We have

$$\mu_X = \frac{\Sigma X}{n_X} \quad \text{and} \quad \mu_Y = \frac{\Sigma Y}{n_Y}$$

For samples, we write

$$m_X = \frac{\Sigma X}{n_X}$$

For example, the president of the local Planned Parenthood Association has four boys, ages 18, 11, 15, and 9, and three girls, ages 18, 2, and 10. If we let X represent the distribution of boys' ages, and Y represent the distribution of girls' ages, we have

$$n_X = 4 \qquad n_Y = 3$$

X	Y
18	18
11	2
15	10
9	
$\Sigma X = 53$	$\Sigma Y = 30$

$$\mu_X = \frac{\Sigma X}{n_X} = \frac{53}{4} = 13.25$$

$$\mu_Y = \frac{\Sigma Y}{n_Y} = \frac{30}{3} = 10$$

SHORTCUTS IN COMPUTING THE MEAN

If you want to compute the mean for numbers which are large and you do not have access to a calculator, you can find the mean by (1) selecting some convenient number, (2) subtracting this number from each raw score, (3) finding the mean of these smaller numbers, and (4) adding back the number you originally subtracted. This technique is called **coding.**

EXAMPLE 2-1　　Find the mean of 101, 102, and 106.

SOLUTION　*Without coding*

X
101
102
106
$\Sigma X = 309$

$$m = \frac{309}{3} = 103$$

With coding　Suppose we decide to subtract 100 from each raw score.

X	$X - 100$
101	1
102	2
106	6
	$\Sigma(X - 100) = 9$

$$\frac{\Sigma(X - 100)}{3} = \frac{9}{3} = 3$$

The mean of the original list is found by adding back the number 100, which we had subtracted from each term, to 3, which is the mean of the coded distribution

■　$m = 3 + 100 = 103$

EXERCISES

2-1　Given Y: 2, 3, 4, 5, 6, 7, 8, calculate each of the following quantities:

(a) $\Sigma Y =$

(b) $\Sigma Y^2 =$

(c) $(\Sigma Y)^2 =$

(d) $\dfrac{\Sigma Y}{n} =$

(e) $\Sigma(Y - 2) =$

(f) $\Sigma(Y - 5) =$

(g) $\dfrac{\Sigma(Y - 5)^2}{n - 1} =$

(h) $\dfrac{\Sigma Y^2 - \dfrac{(\Sigma Y)^2}{n}}{n - 1} =$

2-2　Repeat Exercise 2-1 with Y: 3, 4, 5, 6, 7, 8, 9.

2-3　Given a sample of six values, X: 4, 4, 3, 0, -1, 2, calculate each of the following quantities:

(a) ΣX (b) m (c) $\Sigma(X - m)$

(d) $(\Sigma X)^2$ (e) ΣX^2 (f) $\dfrac{\Sigma X^2 - \dfrac{(\Sigma X)^2}{n}}{n - 1}$

2-4 Repeat Exercise 2-3 with X: 3, 3, 2, -1, -2, 1.

2-5 Given a population of five values, X: 2, 7, 6, 11, 0, calculate each of these quantities:

(a) $\dfrac{\Sigma X^2 - \dfrac{(\Sigma X)^2}{n}}{n}$ (b) $\dfrac{\Sigma(X - \mu)^2}{n}$

2-6 Repeat Exercise 2-5 with X: 4, 14, 12, 22, 0.

2-7 A family had kept track of the age at death of its members over several generations. The ages are 72, 68, 0, 67, 45, 7, 70, 68, 72, 66, 70. Compute the mean, median, and mode ages, and decide which you think is most meaningful.

2-8 Salaries in a mathematics department were as follows: four people at \$9,000, six at \$10,000, two at \$15,000, and one at \$22,000. Compute the mean, median, and mode salaries. Which seems most meaningful?

2-9 Find the mean, median, and mode of the following grade-point averages: 2.9, 3.1, 3.4, and 3.8.

2-10 Illustrate by an example the sentence from this chapter which states "the median is less affected by extremely large or extremely small values, while the mean is affected by every score."

2-11 Write three different distributions where each distribution contains five numbers and has a mean equal to 70. How many such distributions of numbers is it possible to find?

2-12 A student trying Exercise 2-11 took for the first four numbers in one distribution 0, 1, 2, 3. Can he still complete the distribution so that it will have a mean of 70? Could he have started off with *any* four numbers?

2-13 Find the mean of the following arithmetic test grades: 70, 75, 80, 81, 82, 83, 85, 85, 86, 86, 86, 89, 90, 90, 91, 92, 94, and 95. *Hint:* Code by subtracting 85 from each raw score.

2-14 Would there have been any difference in the answer to Exercise 2-13 if we had coded by subtracting 84? Try it.

2-15 A mistake was made in grading the arithmetic papers in Exercise 2-13, and each student is entitled to five more credits than was given. Correct the value for the mean found in the answer to Exercise 2-13.

2-16 The cost of a share of stock in the industrial contractor, Pollution Industrial Group, over the past month has been 587, 588, 588, 590, 593, 597, 597, 600, 601, 599, 598, 598, 597, 599, 600, 603, 605, 605, 604, 607, 605, and 607. Find the mean cost of a share of stock. *Suggestion:* Code or use calculator.

2-17 Find the mean weight of a group of football players whose weights are 307, 225, 256, 301, 353, 227, 256, 210, 302, 269, 285, 291, 287, 253, 271, 298, 303, and 156. *Suggestion:* Code or use calculator.

2-18 The mode was defined to be the most frequent number appearing in a distribution. Some distributions may have more than one mode.

Find the modes in each of the following:
(a) 5, 3, 7, 3, 8, 5, 7, 1, 3, 6, 2, 8, 7
(b) 2, 0, 3, 3, 0, 5, 2, 6, 0, 7, −1, 2, 3
(c) 1, 5, 9, 7

2-19 Some authors prefer the symbol $\overline{X}$ (read X bar) for the mean of a distribution labeled X.
(a) If $X = 3, 5, 0, −1, 7, 2, 4, −2, 8$, find $\overline{X}$.
(b) If $Y = 2, 4, 6, 8, 10$ find $\overline{Y}$.
(c) Write a formula for $\overline{X}$, when you have a distribution of n numbers.

MEASURES OF VARIABILITY

Suppose you are planning to go on a Caribbean cruise during your spring vacation. A travel agent tells you that there are three possible cruises, and that the mean ages of the passengers on each ship are 20, 29, and 41. Which cruise will you select? Which one will your mother select?

Did you pick the ship with a mean age of 20? 29? Or 41? After you have made your choice, look at Table 2-1 for a detailed listing of the passengers' ages. Having seen the passenger lists, would you like to change your selection? You can see that the mean does not accurately reflect the distribution of ages in ships 1 and 2. We need a measure that will indicate whether the numbers in a distribution are close together or far apart. Such a measure is called a **measure of variability, scatter,** or **spread.** Ideally such a measure should be large if the raw scores are spread out and small if they are close together.

One simple measure of variability is the **range.** The range is the difference between the largest number in the distribution and the smallest number. Thus in ship 1 the range is $62 − 2 = 60$ years, in ship 2 the range is $52 − 19 = 33$ years, and in ship 3 the range is $43 − 39 = 4$ years.

For some "everyday" problems, which you might have to solve on an intuitive basis, the range serves very well as the measure of variability. For many more "technical" problems, especially the type we will be doing later in this book, there is another measure of variability that is useful. It is called the **standard deviation.**

To illustrate the concept of standard deviation, let us consider two small populations. Two students, David and Laura, have the same mean grade in algebra, 70. David's grades were 67, 70, 72, and 71. Laura's grades were 100, 62, and 48. Laura's grades are spread out while David's are close together. One way this can be seen clearly is to look at the **deviations from the mean.** The deviation of a score from the mean is found by subtracting the mean from that score (Table 2-2).

A positive sign on a deviation tells us that a grade is above the mean, while a negative sign indicates that it is below the mean. A zero deviation indicates that a particular grade equals the mean. Note that the deviations in David's grades are smaller than those in Laura's grades. This is a result of the fact that David's grades are less scattered than Laura's.

If you compute the mean of the deviations for David's grades you will find that it is zero: $(−3 + 0 + 2 + 1)/4 = 0/4 = 0$. If you compute the mean of the deviations of Laura's grades you will see that it also is zero. In fact it is true for any distribution that the mean of the deviations is zero.

Table 2-1	Ages of the Passengers on Each Cruise Ship		
	ship 1	ship 2	ship 3
	2	19	39
	3	20	39
	4	20	39
	5	21	39
	8	22	41
	9	23	41
	9	23	41
	10	24	41
	40	25	43
	44	49	43
	44	50	43
	62	52	43
	sum = 240	sum = 348	sum = 492
	$\mu = \dfrac{240}{12} = 20$	$\mu = \dfrac{348}{12} = 29$	$\mu = \dfrac{492}{12} = 41$

Recall that we are trying to introduce a new measure of variability called the standard deviation. We want the standard deviation to be representative of the deviations. You might think to use the mean of the deviations as the representative deviation, but we have just mentioned that it is always zero, no matter how much variability there is in the distribution. So statisticians have developed a procedure that is not immediately obvious. First the *squares* of the deviations are obtained, and their mean is found. For example, the deviations in David's grades were $-3, 0, 2$, and 1, and so the squared deviations are $9, 0, 4$, and 1. The mean of the squared deviations is $(9 + 0 + 4 + 1)/4 = 14/4 = 3.5$. This number, the mean of the squared deviations $(9 + 0 + 4 + 1)/4$, can be used as a measure of variability. It is called the **variance** of David's grades. Chapter 15 deals with several situations where the variance is the easiest measure of variability to use.

You will notice, though, that in this process we have squared all the original deviations, so that the variance is representative of the *squares* of the deviations. To get a number representative of the original deviations, we take the square root of the variance. This final number is called the *standard deviation.* In the case of David's grades, since the variance was 3.5, the standard deviation was $\sqrt{3.5} = 1.87$. Since the deviations from the mean were between 0 and 3 units, 1.87 is reasonable as a representation of the deviations. The standard deviation is always in the same units as the original raw scores. In this case the standard deviation is

Table 2-2	David's grades X	mean μ	deviations from the mean $X - \mu$		Laura's grades Y	mean μ	deviations from the mean $Y - \mu$
	67	70	-3		100	70	$+30$
	70	70	0		62	70	-8
	72	70	$+2$		48	70	-22
	71	70	$+1$				

1.87 grade points. If we have an example where the raw scores are in pounds, then the standard deviation is also in pounds.

Let us now calculate the standard deviation of Laura's grades (Table 2-3).

Table 2-3	Laura's grades X	deviation from the mean of 70 $X - \mu$	squared deviation $(X - \mu)^2$
	100	30	900
	62	−8	64
	48	−22	484
	$\Sigma X = 210$	$\Sigma(X - \mu) = 0$	$\Sigma(X - \mu)^2 = 1{,}448$

Variance = mean of the squared deviations

$$= \frac{\Sigma(X - \mu)^2}{n} = \frac{1{,}448}{3} = 482.67$$

Standard deviation = square root of variance

$$= \sqrt{\frac{\Sigma(X - \mu)^2}{n}} = \sqrt{482.67} = 21.98$$

We use the Greek letter for lowercase s, which is σ (read: sigma), to represent the standard deviation. Therefore σ^2 represents the variance.

Thus $\qquad \sigma^2 = \dfrac{\Sigma(X - \mu)^2}{n}$

and $\qquad \sigma = \sqrt{\dfrac{\Sigma(X - \mu)^2}{n}}$

In most statistical applications we do not know all the data for the population. We usually have only a sample of these data. A common problem for the statistician is to estimate the standard deviation of the population from the sample data. The formula given above for σ gives the standard deviation of the population. It is used when you have *all the population data.* However, when you wish to estimate σ using *only sample data,* the formula must be adjusted. The estimate is denoted by s, and the formula is

$$s = \sqrt{\frac{\Sigma(X - m)^2}{n - 1}}$$

Using $n - 1$ instead of n gives a larger value. This compensates for the fact that estimates which use n in the formula tend to be too small because there is usually less variability in a sample than there is in the whole population. Since s is an estimate of σ based on sample data, s is a statistic, while σ is a parameter. When no confusion will result, statisticians sometimes refer to s as the standard deviation, even though it is only an estimate. The estimate for the variance is given by

$$s^2 = \frac{\Sigma(X - m)^2}{n - 1}$$

A COMPUTATIONAL FORMULA FOR s

It turns out that in practice the above formula for s is often awkward to use because of all the subtractions. A second, more convenient formula is

$$s = \sqrt{\frac{\Sigma X^2 - \frac{(\Sigma X)^2}{n}}{n - 1}}$$

This formula produces the same answer as the previous one. Let us illustrate this by computing s both ways for the sample data 1, 8, 0, 3, 9.

Using $\quad s = \sqrt{\frac{\Sigma(X - m)^2}{n - 1}}$

we need m, $X - m$, and $\Sigma(X - m)^2$

$$m = \frac{\Sigma X}{n} = \frac{21}{5} = 4.2$$

X	$X - 4.2$	$(X - 4.2)^2$
1	−3.2	10.24
8	3.8	14.44
0	−4.2	17.64
3	−1.2	1.44
9	4.8	23.04
$\Sigma X = 21$		$\Sigma(X - 4.2)^2 = 66.8$

$$s = \sqrt{\frac{66.8}{4}} = \sqrt{16.7} = 4.09$$

Using

$$s = \sqrt{\frac{\Sigma X^2 - \frac{(\Sigma X)^2}{n}}{n - 1}}$$

we need ΣX and ΣX^2

X	X^2
1	1
8	64
0	0
3	9
9	81
$\Sigma X = 21$	$\Sigma X^2 = 155$

$$s = \sqrt{\frac{155 - \frac{(21)^2}{5}}{4}}$$

$$= \sqrt{\frac{155 - \frac{441}{5}}{4}}$$

$$= \sqrt{\frac{155 - 88.2}{4}}$$

$$= \sqrt{\frac{66.8}{4}} = \sqrt{16.7}$$

$$= 4.09$$

The corresponding computational formula for the variance is

$$s^2 = \frac{\Sigma X^2 - \frac{(\Sigma X)^2}{n}}{n - 1}$$

A SHORTCUT IN COMPUTING s

In Example 2-1 we saw that coding simplified the computation of the mean of 101, 102, and 106. We subtracted 100 from each raw score and found the mean of 1, 2, and 6 to be 3, which told us that the mean of the original list was 103. Let us compute s for the raw scores X and for the coded list Y and compare. Recall that $m_X = 103$ and $m_Y = 3$.

raw scores X	$X - m_x$	$(X - m_x)^2$
106	3	9
102	-1	1
101	-2	4
$\Sigma X = 309$		$\Sigma(X - m_x)^2 = 14$

coded list Y	$Y - m_Y$	$(Y - m_Y)^2$
6	3	9
2	-1	1
1	-2	4
$\Sigma Y = 9$		$\Sigma(Y - m_Y)^2 = 14$

$$s_X = \sqrt{\frac{\Sigma(X - m_X)^2}{n - 1}} \qquad s_Y = \sqrt{\frac{\Sigma(Y - m_Y)^2}{n - 1}}$$

$$= \sqrt{\frac{14}{2}} = \sqrt{7} = 2.65 \qquad = \sqrt{\frac{14}{2}} = \sqrt{7} = 2.65$$

Since the deviations from the mean are the same for both distributions, s is also the same for both distributions.

Thus we see that coding by subtracting a constant affected the mean but not s. For example, suppose we had coded a distribution of ages by subtracting 25 from each age and then found the mean of these coded ages to be 3 and s to be 7. Therefore, the mean of the original set of ages is $25 + 3 = 28$, while s is still 7.

EXAMPLE 2-2   Find s for 3,841, 3,847, 3,847, 3,840, 3,864.

SOLUTION   If we do not code, this will involve very large numbers such as $(3,841)^2$. In order to work with smaller numbers we will code. We decide to subtract 3,847 from each number. (This results in a coded distribution containing two zeros.)

X	$X - 3,847$
3,841	-6
3,847	0
3,847	0
3,840	-7
3,864	+17

We will call this coded list Y, and we now find s for the distribution Y.

We have already mentioned one measure of an individual score, the *raw score*. Another very important measure of an individual's rank in a population is called the **z score.** The z score measures how many standard deviations a raw score is from the mean. In the example above, $\mu = 83$ and $\sigma = 5$. Therefore, the z score corresponding to 88 (written z_{88}) is $+1$, since 88 is 5 units (1 standard deviation) above the mean; $z_{73} = -2$, since 73 is 10 units (2 standard deviations) below the mean.

A formula for finding the z score corresponding to a particular raw score X is

$$z_{\text{corresponding to a raw score}} = \frac{\text{raw score} - \text{mean}}{\text{standard deviation}}$$

Represented in symbols, this is

$$z_X = \frac{X - \mu}{\sigma}$$

EXAMPLE 2-3 Mike is in the above class. His grade was 69. Find the z score for his grade.

SOLUTION We simply substitute into the formula $X = 69$, $\mu = 83$, and $\sigma = 5$

$$z_{69} = \frac{69 - 83}{5} = \frac{-14}{5} = -2.8$$

This tells us that Mike's score was below average by almost 3 standard deviations. In general, you will see that this is quite a lot below average.

■

If in the above class Beverly had a z score of $+2$, what was her score on the test? Since the standard deviation was 5 and the mean 83, 2 standard deviations above the mean would be $83 + 2(5) = 93$. A formula for the raw score corresponding to a particular z score is

Raw score$_{\text{corresponding to a z score}}$ = mean + z score (standard deviation)

which is equivalent to the last formula. In symbols, this is

$$X_z = \mu + z\sigma$$

If David had a z score of $-.7$, then his raw score is

$$83 + (-.7)(5) = 83 - 3.5 = 79.5$$

EXAMPLE 2-4 If we are using s as an estimate of σ, these two formulas become

$$z_X = \frac{X - \mu}{s} \quad \text{and} \quad X = \mu + zs$$

Suppose someone claimed that the mean depth of successful oil-well drillings is 2,500 feet. If you estimate the standard deviation of these depths from some sample data and you get $s = 100$ feet, find the z score corresponding to a depth of 2,250 feet.

SOLUTION $z_{2,250} = \dfrac{X - \mu}{s} = \dfrac{2,250 - 2,500}{100} = -\dfrac{250}{100} = -2.5$

Similarly, to find the depth that corresponds to a z score of 1.65 we have

■ $X = \mu + zs = 2,500 + 1.65(100) = 2,665$ feet

RELATIVE STANDING VIA z SCORE

Let us consider two test grades you might receive. Suppose you received an 85 in English and a 65 in physics. Clearly you would rather receive a raw score of 85 than a raw score of 65, but a second consideration is how well you did relative to the other students in the class. Suppose we tell you that the mean in English was 70 and the mean in physics was 50. Thus, in both classes you scored 15 points above the mean. Does this mean that, relatively speaking, you did the same in both classes? The answer is no; the number of points above or below the mean is insufficient information to give you a rating relative to your position in the class, as you can see from the class scores given in Table 2-4.

Table 2-4

English	physics
100	65 (your score)
99	57
98	55
85 (your score)	53
73	50
67	49
60 (Alice's score)	47
53	44
45	44 (Alice's score)
20	36
mean = 70	mean = 50
$s = 26.4$	$s = 8.06$

We can see from this table that although you scored 15 points above the mean in both classes, when compared to the other students you did better in physics than in English, since your physics grade was the top in the class while three students scored higher than you did on the English test.

In order to see how well you did compared to the rest of the class you can use z scores. In English your z score is

$$z = \frac{85 - 70}{26.4} = \frac{15}{26.4} = .57$$

In physics your z score is

$$z = \frac{65 - 50}{8.06} = \frac{15}{8.06} = 1.86$$

Thus you see that even though you scored 15 points above the class average in both subjects, your physics score was relatively better.

EXAMPLE 2-5 If Alice scored 60 in English and 44 in physics, which was a better grade relative to the class?

SOLUTION In English, Alice's z score is

$$z = \frac{60 - 70}{26.4} = \frac{-10}{26.4} = -.38$$

In physics, Alice's z score is

$$z = \frac{44 - 50}{8.06} = \frac{-6}{8.06} = -.74$$

Since the 60 in English had the higher z score ($-.38$) than the 44 in physics (z score of $-.74$), Alice did better, relatively speaking, in the English ■ class.

PERCENTILE RANK

Another measure of an individual's position in a population is the percentile rank. This is primarily used for large populations. In a small population one would simply use ordinary rankings, such as "fifth out of nine." Essentially, the **percentile rank** of a raw score tells us the percentage of the distribution which is *below* that raw score. Consider a person whose raw score has percentile rank 75. Approximately 75 percent, or three-fourths, of the population scored below this individual.

In a large distribution, if a raw score of 72 has a percentile rank of 80, then 80 percent of the scores are below 72 and 20 percent are above. Now consider the following example. In a distribution of weights of babies, 70 percent of the babies weighed less than Steven, 10 percent weighed the same as Steven, and 20 percent weighed more than Steven. Since 70 percent weighed less than Steven and 20 percent weighed more than Steven, his percentile rank should be between 70 and 80. We will use 75, the value that is halfway between 70 and 80. You can find the percentile rank of a raw score directly by finding the percent of scores which are below the given score and adding one-half the percent of scores which are the same as that raw score. In our example, 70 percent weighed less than Steven and 10 percent weighed the same as Steven, so the percentile rank of Steven's weight = 70 + ½(10) = 75.

Suppose in a class of 50 Andrea's score on an aptitude test was 603. She finds that 6 grades are above hers, 3 (including Andrea's) are the same as hers, and 41 are below her grade. Thus 41/50 = .82 = 82 percent. scored below Andrea's grade and 3/50 = 6 percent are at Andrea's grade. The percentile rank of Andrea's grade is 82 + ½(6) = 85.

We can express the procedure for finding the percentile rank with a formula. Let B equal the number of raw scores *below* a particular score X. Let E be the number of scores equal to X including X itself. Let n be the total number of raw scores. Then the percentile rank (PR) of X is

$$PR_X = \frac{B + \frac{1}{2}E}{n}(100)$$

In the preceding example, the percentile rank of Andrea's grade is

$$PR_{603} = \frac{41 + \frac{1}{2}(3)}{50}(100) = \frac{41 + 1.5}{50}(100) = \frac{42.5}{50}(100) = 85$$

A grade which has a percentile rank of 85 is said to be "at the 85th percentile." We will write P_{85} (read: P sub 85) to indicate the raw score which is at the 85th percentile. Thus, in the example that we just did Andrea's grade would be at the 85th percentile, and would be symbolized by P_{85}. The idea here is that the percentile *rank* is a value from 0 to 100, while a percentile can be any raw score. Thus we can write both $PR_{603} = 85$ and $P_{85} = 603$.

Generally, we round off percentile ranks to the nearest whole number. For example, in a distribution of 150 numbers, suppose that 97 of the numbers are below 700 and that 11 numbers are equal to 700. We wish to find the percentile rank of 700.

$$PR_{700} = \frac{B + \frac{1}{2}E}{n}(100) = \frac{97 + \frac{1}{2}(11)}{150}(100) = .683(100) = 68.3$$

So we say that the percentile rank of 700 is 68, or $PR_{700} = 68$. We also say that 700 is at the 68th percentile, or $P_{68} = 700$. When you read, for example, that your SAT score is at the 82nd percentile, this means that about 82 percent of those taking the examination scored lower than you.

EXAMPLE 2-6

Refer to the data in Table 2-4. Find the percentile ranks for your grades in English and physics. Find the percentile ranks for Alice's grades.

SOLUTION

The percentile rank for your 85 in English is given by

$$PR_{85} = \frac{6 + (.5)(1)}{10}(100) = 65$$

and the percentile rank for your 65 in physics is given by

$$PR_{65} = \frac{9 + (.5)(1)}{10}(100) = 95$$

The percentile rank for Alice's 60 in English is given by

$$PR_{60} = \frac{3 + (.5)(1)}{10}(100) = 35$$

and the percentile rank for Alice's 44 in physics is given by

$$PR_{44} = \frac{1 + (.5)(2)}{10}(100) = 20$$

We summarize these results in Table 2-5.

Table 2-5

	grade	z score	percentile rank
English, you	85	.57	65
physics, you	65	1.86	95
English, Alice	60	−.38	35
physics, Alice	44	−.74	20

2-56 In a large distribution of ages at Golden Vista Nursing Home, the percentile rank of age 72 is 50. True or false:

(a) The median age in this population is about 72.

(b) The mean age in this population is about 72.

2-57 In a given year the mean length of American-made cars was 171 inches and the standard deviation was 5 inches.

(a) Find the z scores for cars with lengths 169 inches, 171 inches, and 180 inches.

(b) Three models of one manufacturer had z scores of $-1, 0,$ and .3. Find the lengths of these models.

(c) All Colonel Motors cars measured within 2 standard deviations of the mean.

    (1) James claims his CM car was 185 inches long. Why is this not possible?

    (2) What is the maximum possible length of James's car?

    (3) What is the minimum possible length of James's car?

2-58 Below are given student Pentak's scores on some standard exams. Also given are other statistics for the exams.

test	mean	standard deviation	Pentak's score
math	47.2	10.4	83
verbal	64.6	8.3	71
geography	74.5	11.7	72

(a) Transform each of Pentak's test scores to a z score.

(b) On which test did Pentak stand relatively highest? Relatively lowest?

2-59 A doctor collects the heights, weights, and blood pressures of a large group of people called a *control group,* and then computes the three means and the three standard deviations. After taking your height, weight, and blood pressure, the doctor computes your z scores with regard to the control group. They are: height; $z = 2.1$, weight; $z = -1.3$, and blood pressure; $z = .003$. Interpret these results.

2-60(a) Sam Safety took the National Safe Driver Test. Of the 120,000 people who took the test there were 100,000 who scored lower than Sam and 2,400 who scored the same as Sam. Find the percentile rank for Sam's score.

(b) John Hasty took the National Safe Driver Test. Of 120,000 people who took the test there were 18,000 who scored lower than John and 1,000 who scored the same as John. Find the percentile rank for John's score.

2-61 According to the statistics of the Quick and Easy Data Company, family incomes in Nowso County have the following percentile ranks:

$P_{25} = \$4,000 \qquad P_{50} = \$6,800$

$P_{75} = \$10,200 \qquad P_{90} = \$14,500$

State approximately the percentage of the families which earn

(a) Less than $4,000   (b) Less than $6,800

(c) Less than $10,200   (d) Less than $14,500

(e) More than $14,500   (f) Between $4,000 and $10,200

2-62 There are about 500,000 families in Nowso County. Using the fig-

ures from Exercise 2-61, state approximately *how many* families earn
(a) Less than $4,000 (b) Less than $6,800
(c) Less than $10,200 (d) Less than $14,500
(e) More than $14,500 (f) Between $4,000 and $10,200
2-63 A test of susceptibility to photographic stimuli was given to 500 subscribers to *Sportfellow* magazine. Some results are tabled below.

test score	68	84	100	116	132	148
z score	-2	-1	0	1	2	3
percentile rank	2	16	30	50	98	99

By inspection of the table is it possible to tell:
(a) What the value of n is? Why?
(b) What the mean test score was? Why?
(c) What the median test score was? Why?
(d) What percentage of the readers scored below 68? Between 84 and 116?
(e) What the standard deviation for the distribution is?
(f) What test score would be 1.5 standard deviations above the mean?
(g) What test score would be transformed to a z score of -1.2?
2-64 In a study of waste disposal in Nosewer County, it was discovered that the mean amount of garbage was 30 pounds per day per family and the median was 35 pounds per day per family. Which one of the following is true?
(a) Exactly half the families produced 30 pounds or more of garbage.
(b) More than half the families produced 30 pounds or more of garbage.
(c) Less than half produced 30 pounds or more of garbage.
2-65 On an important exam Phil's z score was negative. He claimed that his score had a percentile rank of 60. Can this happen?
2-66 If exactly 50 percent of the scores in a population are below 70, then which of these are correct?
(a) $PR_{50} = 70$ (b) $P_{50} = 70$
(c) $PR_{70} = 50$ (d) $P_{70} = 50$
2-67 Danny Kazort is a demolition expert. He has been able to knock down 75 apartment houses at an average time of 3.5 weeks per house, with a standard deviation of 4 days. One tricky job took 5 weeks. What z score would that job have?
2-68 Bess scored 87 on her typing exam and 78 on her physics exam, yet the z score for the 87 was 0 and the z score for the 78 was 2. Explain how this can be and what it means. With relation to her fellow students, is she a better typing student or a better physics student?
2-69 Show that $X = \mu + z\sigma$ and $z = \dfrac{X - \mu}{\sigma}$ are equivalent formulas by solving $X = \mu + z\sigma$ for z.

2-70 Show that $\sqrt{\dfrac{\Sigma(X - m)^2}{n - 1}}$ is equivalent to $\sqrt{\dfrac{\Sigma X^2 - \dfrac{(\Sigma X)^2}{n}}{n - 1}}$. Hint:

Show $\Sigma(X - m)^2 = \Sigma X^2 - \dfrac{(\Sigma X)^2}{n}$. Recall that $m = \dfrac{\Sigma X}{n}$.

The Histogram

3

Quite often when a collection of statistical data is gathered, it must be organized in some way before much sense can be made of it. Probably the most common way to organize statistical data is to graph them. This gives "shape" to the data and may make certain trends or patterns in the data very clear. There are many kinds of graphs that are used to display data, but for the work covered in this text, we shall be especially interested in the type of graph called a **histogram.** The histogram is related to the simple type of graph called a *bar graph.*

Let us start with an illustration of a bar graph. George Stephen, a famous mathematician, gets bored waiting for the bus each morning. He decides to record the length of time, correct to the nearest minute, that the bus is late each day. The results for the past 30 days are pictured in Fig. 3-1.

Some of the things that you can learn very quickly from this bar graph are, for instance, that the bus was 3 minutes late on 8 days, and that it was never more than 10 minutes late. In a bar graph you see that the frequency with which various events occur is reflected in the *heights* of the bars that represent these events. For example, in the graph in Fig. 3-1, the bar over a lateness of 3 minutes is highest because that event happened more often than any other. Another example is that the bar over the lateness of 3 minutes is *twice as high* as the bar over the lateness of 4 minutes, because a 3-minute lateness happened *twice as often as* a 4-minute lateness.

You will see later on in the text that many questions will be answered by referring to the frequency with which various events occur. Sometimes it will be more convenient to have a graph that indicates frequency as an

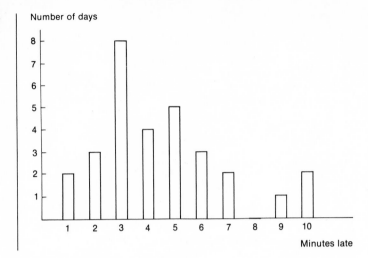

Figure 3-1

Number of days

Minutes late

area rather than as a height. One example we will soon be using is a graph called a *normal curve,* which is illustrated in Fig. 3-2. Here we have graphed a distribution of IQs and shaded the area of the graph representing the proportion of IQs from 115 to 120. The main purpose of this

Figure 3-2

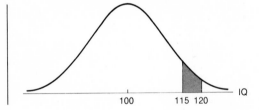

chapter therefore will be to introduce some basic ideas about graphs in terms of *area.* This is most easily done through the type of graph called the histogram.

Let us go back to the problem of George Stephen, the bored mathematician, and describe how to construct the histogram associated with the bar graph that we already have in Fig. 3-1. Basically what we do is this:

1. We make every bar wider until the sides of two adjacent bars meet halfway across the gap that separates them. If there is a bar missing, then we imagine that it is there and that its height is 0. When we do this for the bar graph in Fig. 3-1, we end up with the graph in Fig. 3-3. We have labeled the axis at the points where the new bars meet. These points are called **boundaries.** Notice that the first and last bars have been widened *as if* there were other bars of height 0 next to them. Also notice that the bars over 7 and the 9 have been widened *as if* there were a bar of height 0 over the 8.

2. For clarity, we then remove the original bars. The result is shown in Fig. 3-4. This graph is the histogram.

Figure 3-3

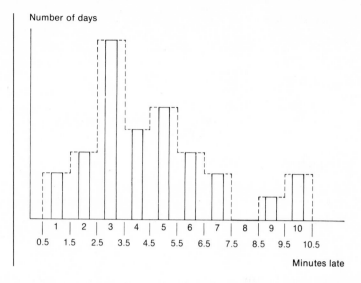

Number of days

Minutes late

Figure 3-4

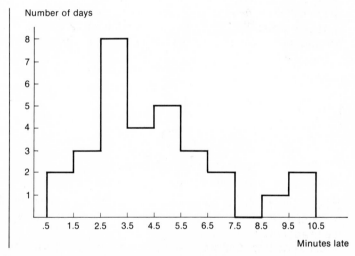

Number of days

Minutes late

Some vocabulary used to describe a histogram:

1. A piece of the horizontal axis between two consecutive boundaries is called an **interval.** For example, the piece from .5 to 1.5 is called the interval from .5 to 1.5.

2. The height of the graph over an interval tells you how many events occurred in that interval. This number is called the **frequency** of that interval. For example, the frequency for the interval from .5 to 1.5 is 2. It means there were 2 days that George's bus was late between .5 and 1.5 minutes.

3. When you subtract the lower boundary of an interval from the upper boundary, the result is called the **width** of the interval. For example, the width of the leftmost interval is 1.5 − .5 = 1. Notice that all the intervals in any one histogram are of the same width. The width of the intervals in a histogram is decided by the person who is making the graph. The width can be small or large depending on the purpose of the graph. For example, George Stephen could have measured how long he waited for the bus to the nearest .5 minute instead of the nearest minute; then he could have chosen intervals of width .5. On the other hand, he might only be interested in waits to the nearest 2 minutes and might set up intervals with width 2.

CONSTRUCTING A HISTOGRAM

In order to construct a histogram we first make a table of the data. Table 3-1 is a table that could have been used to construct the histogram in Fig. 3-4:

Table 3-1	waiting time, minutes	boundaries	frequency
	10	9.5 to 10.5	2
	9	8.5 to 9.5	1
	8	7.5 to 8.5	0
	7	6.5 to 7.5	2
	6	5.5 to 6.5	3
	5	4.5 to 5.5	5
	4	3.5 to 4.5	4
	3	2.5 to 3.5	8
	2	1.5 to 2.5	3
	1	.5 to 1.5	2

Suppose George Stephen had decided to draw a histogram with intervals of width 2. He might have grouped his data as shown in Table 3-2.

Table 3-2	waiting time, minutes	frequency	grouped frequency
	10	2	3
	9	1	
	8	0	2
	7	2	
	6	3	8
	5	5	
	4	4	12
	3	8	
	2	3	5
	1	2	

Then his data would be recorded as in Table 3-3:

**Figure
3-6**

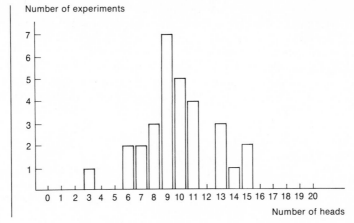

Suppose we want to make a histogram by widening the bars and find-ing the boundaries of the intervals. Figure 3-7 shows what we would get.

**Figure
3-7**

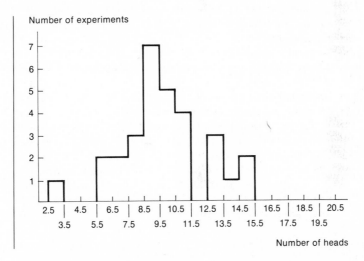

What is the effect of such a graph? It gives the impression that the number of heads is a continuously increasing quantity. It appears, for ex-ample, that in the 7 experiments represented by the interval 8.5 to 9.5 the number of heads could have been anything from 8.5 to 9.5, when in fact the number of heads could only have been exactly 9. On the other hand, the graph still gives an accurate picture of the outcomes of the experi-ments and their frequencies. For example, you can still clearly see that 9 heads was the outcome that occurred more often than any other outcome. And you can still see that no experiment resulted in 12 heads. In other words, the basic *shape* of the distribution is preserved. We will see in Chaps. 5 and 7 that it will be very useful to graph even some count data using histograms because we will be asking questions about areas, which will be easier to answer if we have one continuous graph rather than sepa-

Table 3-5

number of heads in 20 tosses	number of experiments	boundaries	frequency (number of experiments)
15	2	14.5 to 17.5	2
14	1		
13	3	11.5 to 14.5	4
12	0		
11	4		
10	5	8.5 to 11.5	16
9	7		
8	3		
7	2	5.5 to 8.5	7
6	2		
5	0		
4	0	2.5 to 5.5	1
3	1		

rate bars. As long as you keep in mind which kinds of data you have, measure or count data, no confusion will result.

Let us go back to the coin-tossing experiment and discuss drawing a histogram for those data. Two decisions have to be made first: (1) What width should the intervals be? (2) Where should we start the lowest interval? There are no fixed rules for answering these questions. It is up to the person drawing the graph. You might try graphing the same data in histograms of different width intervals to see what the overall effect is. Generally speaking, if the intervals are too wide, you will have too much data lumped together and the trends in the data will be hard to spot. At the other extreme, if the intervals are too narrow the graph will be too spread out and is likely to become very spotty, and once again any trends will be hard to spot. Suppose that for the coin-tossing data we decide to use intervals of width 3, and suppose that we start the lowest interval with a boundary of 2.5. We could tabulate the data as shown in Table 3-5. The histogram for these data would be as shown in Fig. 3-8. You can

Figure 3-8

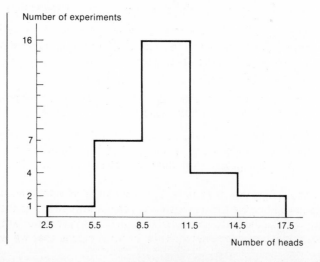

Number of experiments

Number of heads

see that this histogram is similar to Fig. 3-7, but is easier to understand. It shows clearly the idea that in this experiment we often get *around* 10 heads, and that outcomes become more and more rare the further away they are from 10 heads.

What information is contained in a histogram? We note that the percentage of area of the graph which is over any particular interval is equal to the percentage of the outcomes which are in that interval. Let us see why. In Fig. 3-9, we redraw the histogram of Fig. 3-8 and we add some ver-

Figure 3-9

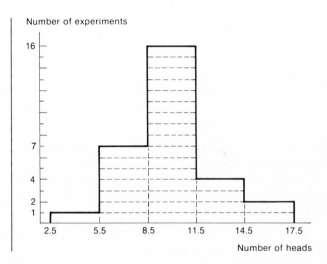

Number of experiments

Number of heads

tical and horizontal lines to show how each of the 30 outcomes is represented by an equal amount of area. This is the key to histograms; *each outcome is represented by an equal amount of area.*

You see for example that there are 16 equal sections in the third interval. This corresponds to the frequency of 16 in Table 3-5. Now let us answer some specific questions to illustrate the principles involved.

1. In what percentage of the experiments was the result from 3 to 5 heads?

Answer From the last two columns of Table 3-5, we see that the frequency for these outcomes was 1. The total number of outcomes was 30. The percentage of outcomes in this interval therefore is 1/30 = .0333 = 3.33 percent.

2. What percentage of the *area* of the histogram corresponds to experiments where the outcome was from 3 to 5 heads?

Answer We see that over this interval is 1 unit of area. The total number of units of area in the graph is 30. Therefore the percentage of area in the graph over this interval is 1/30 = .0333 = 3.33 percent.

You notice then that as far as percentages of outcomes are concerned, the graph and the table contain the same information. In the next chapter you will see that percentages of outcomes are very important in discussing basic ideas of probability. This, in turn, means that we will be able to discuss questions of probability by looking at graphs and measuring percentages of area over various intervals. It will turn out that this is indeed a major tool in statistics because many apparently different questions lead us to graphs of the same shape. So once we know how to measure percentages of area over the intervals of that graph, we will be in good shape to answer those statistical questions. All that remains in this chapter is to show how we can relate what is known about percentages of area over various intervals and the actual overall *shape* of the graph.

PERCENTILE RANK AND z SCORE

In different distributions, the same z score may be associated with different percentile ranks. For example, in a particular distribution of ages of college students, 21 years might correspond to $z = 1$ and have a percentile rank equal to 60, while in a distribution of incomes, \$12,000 might also correspond to $z = 1$, but have a percentile rank equal to 75.

If we know which percentile ranks correspond to the z scores at the boundaries of the intervals, we can draw the histogram. We demonstrate this relationship between z scores, percentile ranks, and area in the following examples.

EXAMPLE 3-1	Draw the histogram for the following data.

z score	percentile rank
2	100
1	70
0	30
−1	20
−2	0

SOLUTION Since the z scores of 0 and 1 have percentile ranks of 30 and 70, respectively, 40 percent of the distribution must lie between $z = 0$ and $z = 1$. In order to draw the histogram, we need to know the area above *each* interval. As we have just shown, you find the area above an interval by subtracting the percentile ranks of its boundaries.

boundaries	percentage of area
1 to 2	$100 - 70 = 30$
0 to 1	$70 - 30 = 40$
−1 to 0	$30 - 20 = 10$
−2 to −1	$20 - 0 = 20$

We can now draw the histogram (Fig. 3-10).

If we had nine z scores instead of five we could draw a more accurate ■ histogram, as indicated in Example 3-2.

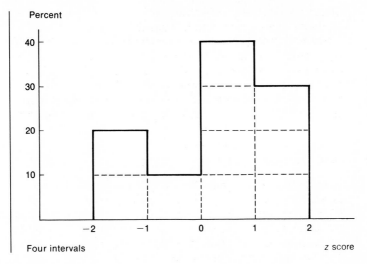

Percent

Four intervals   z score

EXAMPLE 3-2   Draw a histogram similar to the one shown in Example 3-1, given these nine z scores and their corresponding percentile ranks.

z score	percentile rank
2	100
1.5	90
1	70
0.5	55
0	30
−0.5	25
−1	20
−1.5	5
−2	0

SOLUTION   We find the percentage of area for each interval as we did in Example 3-1.

boundaries	percentage of area
1.5 to 2	10
1 to 1.5	20
0.5 to 1	15
0 to 0.5	25
−0.5 to 0	5
−1 to −0.5	5
−1.5 to −1	15
−2 to −1.5	5

■

The histogram for these data is shown in Fig. 3-11.

If we had 100 z scores, we would have a histogram with 99 intervals. This histogram would be difficult to draw and is often approximated by a smooth curve, as shown in Fig. 3-12.

Figure 3-11

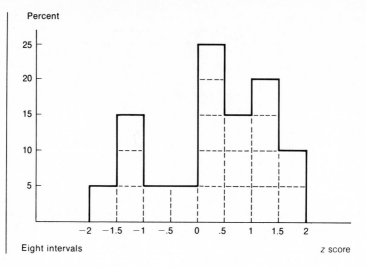

Percent

Eight intervals

z score

Figure 3-12

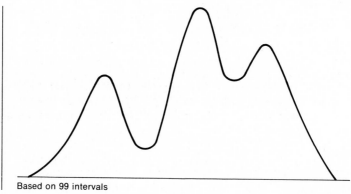

Based on 99 intervals

GLOSSARY

VOCABULARY

1. Bar graph   **2.** Histogram   **3.** Interval
4. Width of intervals   **5.** Boundaries   **6.** Frequency
7. Measure data   **8.** Count data   **9.** Continuous variable
10. Discrete variable

EXERCISES

3-1 Would the following quantities be recorded as count data or measure data?
(a) The number of mangos in a basket
(b) The weight of the mangos in a basket
(c) The average weight of 17 mangos
(d) Lengths of feet
(e) Shoe sizes
3-2 Draw a histogram for the following data:

interval	boundaries	frequency
10	9.5 to 10.5	7
9	8.5 to 9.5	16
8	7.5 to 8.5	15
7	6.5 to 7.5	36
6	5.5 to 6.5	99
5	4.5 to 5.5	150
4	3.5 to 4.5	216
3	2.5 to 3.5	304
2	1.5 to 2.5	401
1	.5 to 1.5	197
0	−.5 to .5	253

3-3 Compute the frequency in each interval by using the data given in Exercise 3-2, and draw the histogram.

interval	boundaries	frequency
9 to 11	8.5 to 11.5	
6 to 8	5.5 to 8.5	
3 to 5	2.5 to 5.5	
0 to 2	−.5 to 2.5	

3-4 For the data shown, compute the boundaries and then draw the histogram. Find the percentage of the distribution in each interval, and notice that it is the same as the percentage of area of the graph at that interval.

interval	frequency	boundaries	percentage
95 to 100	9		
89 to 94	14		
83 to 88	12		
77 to 82	7		
71 to 76	3		
65 to 70	5		
totals =	50		100

3-5 For the data shown answer the same questions as in Exercise 3-4.

interval	frequency	boundaries	percentage
120 to 129	45		
110 to 119	75		
100 to 109	0		
90 to 99	30		
	total = 150		

3-6 An experiment is to roll a die 6 times. This experiment was repeated 20 times, and the number of times that a 2 appeared in each experiment recorded. The results are tabulated by making one tally mark after each 6 tosses of the die.

number of 2s	tally	frequency
6		0
5	/	1
4	/	1
3	///	3
2	//	2
1	///// //	7
0	///// /	6
		20

(a) Draw a histogram for the data in the table.

(b) What is the mean number of times that a 2 appeared on the die?

3-7(a) Toss a coin 5 times, and count the number of times that tails appear. Repeat this experiment 15 times. Record the data as shown below, and draw a histogram for the data. Make one tally mark after each 5 tosses of the coin.

number of tails	tally	frequency
5		
4		
3		
2		
1		
0		
		15

(b) What is the mean number of tails that occurred in 5 tosses?

3-8(a) From this textbook pick 50 pages at random. On each page pick 1 line of text at random, and count the number of e's. For the list of 50 numbers that you get, graph your results in two separate histograms, one with intervals of width 1, the other with intervals of width 2.

(b) What is the mean number of e's per line?

3-9 Select a sample of 50 numbers from the random number table. The numbers should range from 0 to 99. Draw a histogram for your sample using boundaries of −.5, 9.5, 19.5, 29.5, . . . , 99.5. Which of these shapes is closest to your own histogram? Is this what you expected to happen?

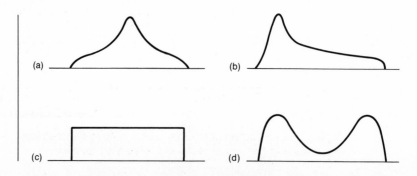

3-10 For the histogram shown below answer these questions:
(a) Find the number of people represented in each interval, and find the total number of people represented in the graph.
(b) Find the percentage of people in each interval.
(c) Find the *percentage* of area of the graph which is in the space over each interval and check that the total of these is 100 percent.

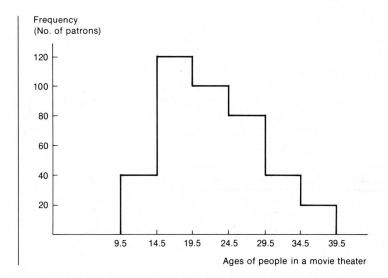

Ages of people in a movie theater

3-11 For the histogram shown below answer the same questions as in Exercise 3-10.

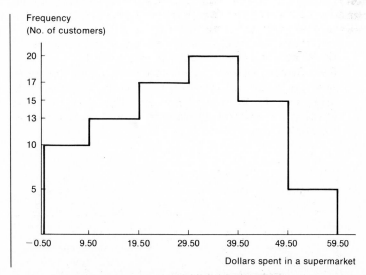

Dollars spent in a supermarket

3-12 The histogram illustrated on page 46 is symmetrical. Some of the z scores and percentile ranks are related as follows:

z score	percentile rank
3	100
2	98
1	84
0	50

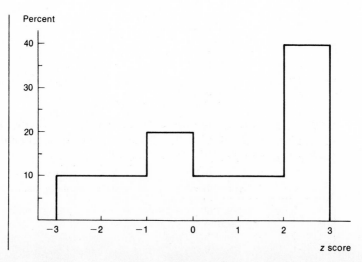

What percentage of the area of the graph is
(a) To the left of 0?   (b) To the left of 1?
(c) To the left of −1?   (d) To the right of 2?
(e) To the right of −2?   (f) Between 0 and 3?
(g) Between 0 and 2?   (h) Between −2 and 0?
(i) Between −1 and 1?   (j) Between −2 and 2?

3-13   Given the following relationship between z scores and percentile ranks and the corresponding histogram, answer Exercise 3-12 (a) to (j).

z score	percentile rank
3	100
2	65
1	45
0	35
−1	30
−2	10
−3	0

If a variable has only 2 possible outcomes *and* if the probabilities of these outcomes do not change for each trial regardless of what has happened on previous trials, then the variable is called a **binomial variable.** We would not consider the variable whose outcomes are rain and no rain to be a binomial variable, since the probability of rain changes each day; however, the result of tossing a coin is a binomial variable since the probabilities for heads and tails remain the same for each toss.

EXAMPLE 5-1
The Knights of Columbus run an annual fund-raising bazaar for a local hospital. A game of chance at the bazaar is set up like this: A spinner has 3 equal areas colored red, green, and blue. A patron bets on 1 color. The wheel is spun, and if it stops with the pointer on the color the patron selected, he wins. Marty decides to play the game twice, betting on red both times. The result of his betting is a binomial variable (win-lose) with 2 trials (since he will bet 2 times). What is the probability that Marty will win

(a) Both times?   (b) Just one time?   (c) Not at all?

We will solve this problem two ways: first, by looking at all the equally likely outcomes as before; and second, by a new approach we will call the "binomial" approach.

SOLUTION   *Method 1*   There are $3 \times 3 = 9$ equally likely possible outcomes, listed below.

outcome	first game	second game	
1	red	red	2 reds
2	red	green	
3	red	blue	1 red
4	blue	red	
5	green	red	
6	blue	green	
7	green	blue	no reds
8	blue	blue	
9	green	green	

(a) The probability that Marty wins twice is $P(2 \text{ wins}) = 1/9$, since there are 9 equally likely outcomes and only 1 of them consists of 2 reds.

(b) The probability that Marty wins exactly 1 time is $P(1 \text{ win}) = 4/9$, because 4 of the outcomes consist of 1 red.

(c) Similarly the probability that Marty does not win at all is $P(0 \text{ wins}) = 4/9$. We can present these probabilities in a histogram as shown in Fig. 5-1.

SOLUTION   *Method 2*   There is an alternative way to compute these probabilities (1/9, 4/9, 4/9) which is often easier. We can analyze the game directly in terms

Figure 5-1

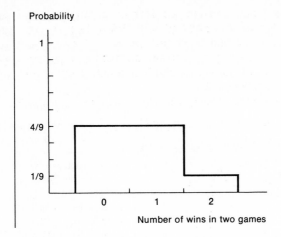

Probability

Number of wins in two games

of win and lose, instead of in terms of color. As far as Marty is concerned, when he plays twice, he will win twice, once, or not at all. There is only one way he can win twice (he must win on both games). There are two ways he can win once (either win on the first game, *or* win on the second game). And there is only one way he can lose twice (he must lose on both games). We summarize this in Table 5-1.

Table 5-1

number of wins	ways the wins can occur	number of different ways the wins can occur
2	WW	1
1	WL, LW	2
0	LL	1

Recall that in a binomial problem there are 2 possible outcomes on each trial. One of these is called a "success" and the other is called a "failure." We are now treating this problem as a binomial problem where "success" is win, "failure" is lose, and a "trial" is one spin. It is standard in this type of problem to let n stand for the number of trials, to let S stand for the number of successes in n trials, and to let $\binom{n}{S}$ stand for the number of different ways S can occur. The symbol $\binom{n}{S}$ is called a **binomial coefficient.** Thus, we can relabel Table 5-1 as shown in Table 5-2.

Table 5-2

S	ways S can occur	$\binom{2}{S}$
2	WW	$\binom{2}{2} = 1$
1	WL, LW	$\binom{2}{1} = 2$
0	LL	$\binom{2}{0} = 1$

lands on J, he spends the evening at Jay's. If the spinner lands on K, he spends the evening at Ken's. There is a 4-day holiday coming up, and so Chastain spins the spinner 4 times to determine his plans. What is the probability that
(a) He will spend 1 evening at Jay's and 3 evenings at Ken's?
(b) He will spend 2 evenings at Jay's and 2 at Ken's?
(c) He will spend all 4 evenings at the same night spot?

5-16 *Suggestion:* Use calculator. Assistant Professor Ratso, a leading experimental psychologist, is in the habit of sending rats through mazes. He predicts that a rat reaching the end of a T-shaped maze is more likely to turn left than right. He believes that the proportion of rats which turn left is .65. If this is true and he sends 6 rats down a maze
(a) What is the probability that 3 will turn left and 3 will turn right? .24
(b) What is the probability that fewer than 5 will turn left? .68
(c) What is the probability that all will turn left or all will turn right? .08

5-17 *Suggestion:* Use calculator. An emergency life support system has 4 batteries. The probability of any battery failing is .01. What is the probability that
(a) None fail? (b) All fail? (c) More than 2 fail?

5-18 A manufacturer claims that 4 of 5 dentists recommend sugarless gum for their patients who chew gum. Assuming that this claim is true, find the probability that in a randomly selected group of 20 dentists, 16 or more will recommend sugarless gum for their patients who chew gum. *Suggestion:* Use calculator.

5-19 Find S if

$$\binom{n}{3} + \binom{n}{4} = \binom{n+1}{S}$$

5-20 Suppose that 2 baseball teams, A and B, are equally matched. The outcome of each game between them can be considered as a random variable, with the probability of team A winning equal to .5.
(a) If they play 6 games, what is the probability that each team will win 3 games?
(b) If they play a series of games where the first team to win 4 games wins the series, what is the probability that team A will win the series *on the seventh game?* *Hint:* Use your answer to part (a).

5-21 Chris and Evonne, 2 tennis players, are playing a game. The probability that Chris wins a point is .8. The game is currently stopped with "advantage Evonne" (she needs 1 point to win). What is the probability that Chris will win the next 3 points consecutively, and thus win the game?

5-22 Peter Pedaler must bike 2 miles to get the Sunday newspaper. If he gets to the store too early, the papers have not arrived. If he gets there too late, they are all sold. He has learned that if he arrives at 8:30 he has

.895 an 85 percent chance of getting the paper. What is the probability that he will get the paper at least 6 of the next 8 Sundays if he shows up at 8:30?

5-23 Chef Victoir knows how to prepare only 2 dishes: Quiche Fillisse and Eggs Ari-Bari. A patron of his restaurant, Mme. Sharonne, has noticed that 63 percent of the time Victoir prepares Eggs Ari-Bari. If Mme. Sharonne enters the restaurant 6 times, what is the probability that she will get 3 meals of each dish?

p^3q^3

Label the raw scores corresponding to the z scores given on the axis on the above graph.

6-11 $ND\ (\mu = .6,\ \sigma = .1)$

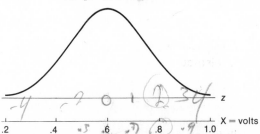

Label the z scores corresponding to the raw scores given on the above graph.

6-12 An electronics technician repeats an experiment many times, each time recording a voltage reading. The technician finds that the collection of readings is approximately normally distributed, with μ about 74 volts and σ about 6 volts.

(a) What percentage of the readings were between 70 and 80 volts?

(b) What is the probability that a reading taken at random will be over 86 volts?

(c) What is the probability that a random reading will be *outside* the range of 69 to 79?

6-13 The results on a certain blood test performed in a medical laboratory are known to be normally distributed, with $\mu = 60$ and $\sigma = 18$.

(a) What percentage of the results are between 40 and 80?

(b) What percentage of the results are between 76 and 78?

(c) What percentage of the results are above 100?

(d) What percentage of the results are below 60?

(e) What percentage of the results are between 78 and 80?

(f) What percentage of the results are outside the "healthy range" of 30 to 90?

(g) What is the probability that a blood sample picked at random will have results in the "healthy range" of 30 to 90?

(h) Which test result has a percentile rank of 5?

6-14 A study was done to see how many hours during the school day high school seniors spend thinking about sex. The results were normally distributed with $\mu = 2.7$ hours and $\sigma = .6$ hour. What percent of these students think about sex

(a) More than 1 hour per school day?

(b) More than 4.5 hours?

(c) Between 2 and 3 hours?

6-15 At an urban hospital the weights of newborn infants are normally distributed, with $\mu = 7$ pounds, 2 ounces, and $\sigma = 15$ ounces. Let X be the weight of a newborn infant which is picked at random. Find the following probabilities:

(a) $P(X \geq 8 \text{ pounds})$

(b) $P(X \leq 5 \text{ pounds, 5 ounces})$

(c) $P(6 \text{ pounds} \leq X \leq 8 \text{ pounds})$

(d) What infant weight is at the 70th percentile? (That is, find P_{70}.)

(e) Let W be a fixed weight. The probability that a baby picked at random weighs less than W is .70. Find W. [That is, find W such that $P(X < W) = .70$.]

(f) Find W such that $P(X < W) = .10$.

6-16 The lifetimes of a certain brand of movie floodlights are normally distributed, with $\mu = 206$ hours and $\sigma = 56$ hours. Let X be the lifetime of a light picked at random. Find the following probabilities:

(a) $P(X \geq 300)$

(b) $P(X \leq 100)$

(c) $P(100 \leq X \leq 300)$

(d) The company guarantees that its light will last at least 120 hours. What percentage of the bulbs do they expect to have to replace under this guarantee?

6-17 It happens that income for junior executives in a large retailing corporation is normally distributed, with $\mu = \$16,400$ and $\sigma = \$1,500$.

(a) There is an unspoken agreement that you have "arrived" if you are in the top 15 percent. What salary must a junior executive earn in order to arrive?

(b) The highest 25 percent get keys to the executive washroom. Victoria earns $17,540. Does she have a key?

(c) Due to a recession the bottom 5 percent may be let go. What salary cuts off the bottom 5 percent?

(d) The top 20 percent go out to lunch. The bottom 30 percent bring their lunch in interoffice envelopes. The remaining 50 percent bring their lunches in attaché cases. Find the two salaries that separate these three categories.

6-18 True or false? For a normal curve

(a) The area between $z = 0$ and $z = 1$ equals the area between $z = 0$ and $z = -1$.

(b) The area between $z = 0$ and $z = 1$ equals the area between $z = 1$ and $z = 2$.

(c) The percentile rank of $z = 1$ equals the percentile rank of $z = -1$.

6-19 The graphs below represent the distribution of incomes in two populations. Discuss any differences between the two distributions.

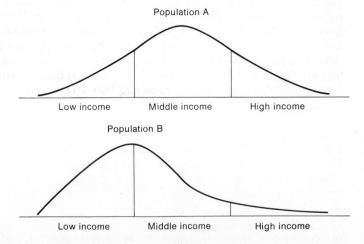

Population A

Low income   Middle income   High income

Population B

Low income   Middle income   High income

$H_a: p > .34$

2. If you suspect that p is less than .34 you would write

$H_a: p < .34$

3. If you don't have any idea whether p is larger or smaller than .34, then you could write $p \neq .34$.

In the first instance, you would only be interested in values of p *greater than* 34 percent, and in the second instance you would only be interested in values of p *less than* 34 percent. These are referred to as **one-tail tests,** since the values of interest are in only *one* direction away from 34 percent. The third instance, however, is referred to as a **two-tail test,** since values far from 34 percent in *either* direction are of interest to you in your experiment.

Note that we have formulated our hypotheses so that the equals sign (=) always appears in the null hypothesis, while either the less than (<) or the greater than (>) signs appear in the alternative hypothesis for a one-tail test. The alternative hypothesis for a two-tail test always contains the not equal sign ($\neq$). The choice of either a one-tail or two-tail test is determined by what the statistician is interested in finding out.

EXAMPLE 8-1 Formulate the null hypothesis and alternative hypothesis for each of the following.

    (a) Is the average life span of a dog more than 13 years?

    (b) Is the proportion of 18-year-old drivers who have accidents the same as the proportion of 26-year-old drivers who have accidents?

    (c) What percentage of people born with Down's syndrome can be taught to read?

    (d) Do teenage girls receive a smaller weekly allowance than teenage boys?

SOLUTION   (a) Let μ = the average life span of dogs

        $H_0: \mu = 13$

        $H_a: \mu > 13$ (one-tail test)

    (b) Let $p_1 = P(\text{18-year-old driver has an accident})$

        $p_2 = P(\text{26-year-old driver has an accident})$

        $H_0: p_1 = p_2$ or $p_1 - p_2 = 0$

        $H_a: p_1 \neq p_2$ or $p_1 - p_2 \neq 0$ (two-tail test)

    (c) This calls for a numerical response, not a hypothesis test.

    (d) Let μ_1 = average weekly allowance for teenage girls

        μ_2 = average weekly allowance for teenage boys

        $H_0: \mu_1 = \mu_2$ or $\mu_1 - \mu_2 = 0$

■ $H_a: \mu_1 < \mu_2$ or $\mu_1 - \mu_2 < 0$ (one-tail test)

EXERCISES

Formulate the two hypotheses where applicable, and decide whether the situation calls for a one-tail or two-tail test.

8-1   Is the average work week in Centerville less than 40 hours?

8-2   Do more than 10 percent of pet owners own goldfish?

8-3   What is the average jail sentence for bank robbery?

8-4   Do 12 percent of the students in your school major in mathematics?

8-5   Is the average height of 6-year-old male horses the same as the average height of 6-year-old female horses?

8-6   What is the average weight of adult baboons?

8-7   Do a greater percentage of 20-year-old females diet than do 20-year-old males?

DECISION RULES

At the beginning of an experiment you should formulate the two opposing hypotheses. Then you should state what evidence will cause you to say that you think the alternative hypothesis is the true one. This statement is called your **decision rule.** When the evidence supports the alternative hypothesis, we say that we "reject the null hypothesis." When the evidence does not support the alternative, we say that we "fail to reject the null hypothesis."

EXAMPLE 8-2   Guildenstern suspects that a certain coin is biased for heads. She decides to test it by tossing it 40 times. Her null hypothesis is H_0: coin is fair, $p = P(\text{heads}) = .5$. H_a: coin is biased for heads, $p > .5$. She reasons that if it is fair, then she should get about 20 heads. She makes the following decision rule: If the 40 tosses produce 25 or more heads, then conclude that the coin is biased in favor of heads. If she lets S stand for the number of heads she will get, then, her decision rule is: If S is more than 25, reject H_0.   ∎

EXAMPLE 8-3   Amir Treifel is testing the null hypothesis that the mean income of sheiks is 1.5 million dollars per year, H_0: $\mu = 1.5$ million dollars; therefore, his alternative hypothesis is H_a: $\mu \neq 1.5$ million dollars. He decides that he will reject his null hypothesis if the mean income for a random sample of sheiks turns out to be either less than 1 million or more than 2 million dollars. If he lets m stand for the mean of his sample, then, briefly, his decision rule is: If m is less than 1 million or more than 2 million dollars, reject H_0.   ∎

STATISTICAL ERRORS

One basic idea which is inseparable from hypothesis testing is that you can almost never have *absolute* proof as to which of the two hypotheses is the true one. For example, in the case of testing a coin to see if it is fair, you must realize that the very definition of "fair coin" makes it impossible to completely test a coin. Recall that a fair coin is one for which the probability of heads on a single toss equals .5. But, "a probability equal to .5" company the benefit of the doubt, and so his decision rule is: If more than 40 trains are late, reject the company's claim.

(a) State the motivated hypothesis.

(b) State the null hypothesis.

(c) Draw the normal curve including the z scale line, the line for the number of successes, and a shaded rejection region.

(d) Find the probability of a Type I error.

(e) If the decision rule were reject H_0 if more than 30 trains are late, find alpha.

(f) If in part (e) you change the number 30 to 50, would alpha increase or decrease?

(g) What decision rule gives $\alpha = .05$?

8-22 The population of Smalltown is 42 percent female. A statistician is trying to establish the claim that more than 42 percent of the Republicans in Smalltown are women. Assuming that sex has nothing to do with political affiliation, then $p = P$(a Republican is female) $= .42$. We select 100 Republicans at random. Using the .01 significance level, find the decision rule for the number of women that must occur in the sample before we reject the assumption. 54

8-23 A newspaper article states that 60 percent of children in the age bracket 1 to 4 who die do so as a result of motor accidents. Doubting that this is true, a public health official gathered information on the cause of death of 30 randomly selected children.

(a) State the motivated hypothesis.

(b) State the null hypothesis.

(c) How many or how few of the 30 deaths should be due to motor vehicles in order to reject the claim of the article at the .01 significance level?

8-24 A manufacturer claims that the mixed nuts that he sells have only 30 percent peanuts. We open a large bag and select 100 nuts at random and find that 36 of them are peanuts. If p does equal .30, what is the probability that 100 nuts picked at random contain 36 or more peanuts? Would you be willing to accuse the manufacturer of a false claim? P=.1151 Possible

8-25 A gambler tosses a *fair* coin 10 times and gets 8 heads. He makes a Type I error in stating that the coin is biased. Find the probability that a fair coin tossed 10 times will produce 8 or more heads.

MORE ABOUT TYPE II ERRORS

(This section may be omitted without loss of continuity.)

We have just discussed the Type I error, which occurs when statistical evidence leads us to reject a null hypothesis when in reality the null hypothesis is true. Recall that the Type II error occurs when the null hypothesis is false but the statistical evidence is not strong enough to indicate it. That is, we have mistakenly failed to reject the null hypothesis. The probability of a Type II error is denoted by β. If the value of p given in the null hypothesis is wrong, then some other particular value of p is correct. Recall that the alternative hypothesis did not give us one specific value of p. Therefore, for each possible value of p, there is a corresponding value of beta.

EXAMPLE 8-7

Suppose you are asked to test a coin. You decide to toss it 60 times, reasoning that if it is fair you will get about 30 heads. You choose for your decision rule: The coin is not fair if the number of heads is less than 26 or more than 34.

If unknown to you, the coin *is* biased and $p = P(\text{heads}) = .6$, then what is the probability that your experiment will yield between 26 and 34 heads anyway? That is, what is the probability that you commit a Type II error?

SOLUTION

$p = .6$

$q = .4$

$n = 60$

$np = 36 > 5$

$nq = 24 > 5$

Therefore, the normal approximation may be used, as shown in Fig. 8-4.

Figure 8-4

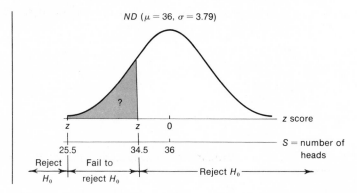

$ND\ (\mu = 36,\ \sigma = 3.79)$

$\mu = 36$

$\sigma = \sqrt{14.4} = 3.79$

$P(26 \le S \le 34)$ is approximately equal to $P(25.5 < S < 34.5)$

$z_{34.5} = \dfrac{34.5 - 36}{3.79} = -.40$

$z_{25.5} = \dfrac{25.5 - 36}{3.79} = -2.77$

Area $(z = -.40) = .3446$

Area $(z = -2.77) = .0028$

■ Therefore $P(26 \le S \le 34) = .3446 - .0028 = .3418$.

The probability is .34, or about 1/3, that we would accept the statement "this coin is fair," when in fact the coin is biased with $p = .6$. We write $\beta = .34$. You see that our decision rule is not very powerful for distinguishing between fair coins and coins biased with $p = .6$. You will notice that we needed the probability $p = .6$ in order to compute beta. We cannot compute a beta until we choose a specific value of p from the alternative hypothesis. We could repeat the calculation above for other values of p,

8-56 The usual dropout rate in the freshman class at Wealth College is 50 percent. A new dean of admissions claims that her policies have lowered the dropout rate because in this year's class of 600 freshmen only 260 dropped out. Test at the .05 level. Do the statistics support her claim? *less 279.8 Rej Ho*

8-57 Using a particular type of bombing mechanism usually results in 70 percent of the bombs being on target. An engineer claims to have invented a more accurate mechanism. The new mechanism is used to drop 100 test bombs on a target; 75 hit the target. Is this enough evidence to establish that the new mechanism is better than the old? Use $\alpha = .05$.

8-58 According to a nationwide poll, only 40 percent of the voters favor a health care bill which would benefit the poor. A certain politician believes that the percentage of voters who favor the bill is higher in his district. A random sample of 30 voters in his district is taken and it is found that 14 support the health care bill. Is this sufficient evidence to support the politician's belief at the .01 significance level? *> 18.24*

8-59 The National Safety Council claims that 25 percent of automobile accidents involve pedestrians. A government statistician feels that the true figure is smaller than 25 percent. A random sample of 36 automobile accidents is selected and it is found that 11 of them involved pedestrians. Is this sufficient evidence to reject the council's claim at the .01 significance level?

8-60 A certain medicine is said to be at least 90 percent effective in giving relief to people with allergic reactions to cats and dogs. Dr. Kay Nyne believes that this claim is incorrect. A random sample of 60 people with such allergies is selected from patients at an allergy clinic. What would you say about the claim at the .05 significance level if 58 people got relief? *≥ 50.17*

8-61 Ralph claims that on Saturday morning TV programs on WW-TV 25 percent of the time is devoted to commercials. A student tests this one Saturday by switching on the TV at 50 randomly chosen times between 7 A.M. and noon. She finds that at 9 of these times she sees a commercial. Does this support Ralph's claim? (Use $\alpha = .05$.)

8-62 It is known that in July on a certain stretch of the northeast coast of the United States 60 percent of the sea gulls are Franklin's gulls. A birdwatcher goes out one morning and spots 80 gulls. He finds that 75 of them are Franklin's gulls. Give two possible explanations for this. Use $\alpha = .01$.

8-63 A professor reads in the paper that 60 percent of all college freshmen are more interested in being popular than in doing well at school. She thinks this is too high, so she goes around and interviews 100 freshmen. She finds that 10 say they are more interested in being popular, while 90 say they are more interested in doing well. Test her results at .05 significance level. Give several interpretations of your results.

8-64 An economics student reads that in his county 35 percent of the employed earn more than $15,000 per year. He wants to see if the claim is accurate so he mails out 500 questionnaires to people chosen at random from the phone book. He gets back 100 replies; 80 of them report incomes of more than $15,000. Give several interpretations of these results. He plans to test at the .05 significance level.

8-65 Under certain conditions the probability is .10 that a tadpole survives to mature into a frog. Now a scientist believes that he has found a way to place vitamins in the frog pond so that more tadpoles will survive. Using this new approach we take a random sample of 98 tadpoles, and test using the .05 significance level. For part (e), if 12 tadpoles survive, state whether you reject or fail to reject the null hypothesis, and explain what that means. If 27 tadpoles survive, state whether you reject or fail to reject the null hypothesis, and explain what that means.

8-66 Stan Sly claims that he can control the tosses of a fair coin. To see if he is correct you take two ordinary coins and give him one. You toss one and ask him to try to toss the same thing, that is, if your toss results in a tail then he is to try to toss a tail also. You repeat this experiment 18 times and Stan succeeds in tossing the same as you 15 times. Is this unusual at the .05 significance level? *12.5*

8-67 With a fair pair of dice, doubles (two numbers the same) should come up about 1/6 of the time. Marsha, who is losing at Monopoly, has rolled 15 doubles in the past 60 rolls. She claims that the dice are biased.
(a) Find the probability that a pair of honest dice could produce 15 or more doubles in 60 rolls.
(b) Is this probability that you found in part (a) alpha, beta, or neither?

8-68 How large a sample size would be needed to use a normal curve approximation to test the hypothesis that 3 percent of left-handed Carpathians have at least one green eye?

FIELD PROJECTS

From this point on we will include suggestions for field projects. We hope that you will be able to do some during the semester. We suggest that you conduct the field project in two stages.
1. Outline clearly what you *intend* to do. State the population or populations that you wish to sample. Describe your intended sampling procedure. Comment on its strengths and weaknesses. State your null and alternative hypotheses, the significance level, and the sample size. If you are going to ask questions of people in your sample, state these questions now *exactly* as you will ask them. If you are going to count something in your sample, state *exactly* what you are looking for. In any case, state how you will handle responses which do not fit into your predetermined categories. Give all this information to your instructor. After the instructor has approved it, then proceed with stage 2.
2. Perform the experiment as approved. Calculate all your data. Submit your results with comments as to the strengths and weaknesses of the project as you actually carried it out.

EXAMPLE OF A FIELD PROJECT

A claim appears in a newspaper that 60 percent of Americans feel that the President is "doing a good job." A student doubts that this percentage is correct in his neighborhood, and so he sets up the following field project.

Stage 1   (a) *Population* All people 16 years old or older who live within 3 blocks of my house.

(b) *Sampling procedure*   There are 30 blocks in this neighbor-hood. I will pick 3 houses at random on each block. I will question only 1 person at each house. If no one is home, I will select another house on that block.

(c) *Questions to be asked*
1. I am doing a survey for my college class in statistics. Would you please answer the following questions?

2. Are you 16 years old or older?

3. Do you think that the President is doing a good job?
I will continue until I get 90 yes responses to both questions 1 and 2, and either a yes or a no response to question 3. Since $np = 90(.6) = 54 > 5$, and $nq = 90(.4) = 36 > 5$, we can use the normal distribution.

(d) H_0: the fraction in my neighborhood is the same as the frac-tion of all Americans, $p = .60$

H_a: the fraction in my neighborhood is not the same as the fraction of all Americans, $p \neq .60$. (two-tail test)

I will use $\alpha = .05$.

(e) *Comments on strengths and weaknesses*   Depending on when I am able to cover the houses I may get more or less working people in my sample.

Stage 2   (Conducted after stage 1 was approved)

(a) I went to 123 homes. In 12 homes no one answered the door. In 3 homes there was no one over 16 present. In 2 homes the answer to question 1 was no. In 4 homes people first answered yes to question 1 but changed their mind after they heard question 3. In 12 homes the person was unde-cided on question 3. The remaining 90 responses were

yes	no
48	42

SOLUTION FOR THIS PROJECT

H_a: percentage in neighborhood is not the same as percentage of all Americans, $p \neq .60$ (two-tail test)

H_0: $p = .60$

$n = 90$

$\alpha = .05$ (two-tail test)

Since $np = 54 > 5$, $nq = 36 > 5$, we can use the normal approximation with $\mu = 54$ and $\sigma = \sqrt{21.6} = 4.6$.

$z_c = \pm 1.96$

The curve for this distribution is shown in Fig. 8-11.

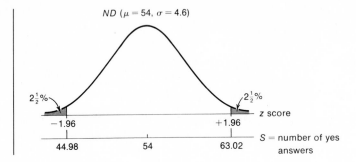

**Figure
8-11**

$ND\ (\mu = 54,\ \sigma = 4.6)$

$2\frac{1}{2}\%$

$2\frac{1}{2}\%$

z score

−1.96

+1.96

S = number of yes answers

44.98

54

63.02

$$S_c = 54 + (\pm 1.96)\ (4.6)$$
$$= 54 \pm 9.02$$
$$= 63.02 \quad \text{and} \quad 44.98$$

Therefore, my decision rule is to reject the null hypothesis if I get more than 63.02 yes answers or fewer than 44.98 yes answers. Since the experimental outcome was 48 yes answers I have failed to reject the null hypothesis. I was unable to show that the null hypothesis was false. Of the 33 nonresponses I have no evidence to indicate that they would differ markedly from the 90 who did respond. If there was a marked difference it might change my conclusion.

SUGGESTED PROJECTS

Outline a one-sample binomial hypothesis test to be performed on some population of your choice. After your instructor has approved your outline, gather your data and perform the hypothesis test.

1. Select a reported fact from a newspaper or some other source as in the preceding example and test it on some population of your own choice.
2. Test a theoretical hypothesis concerning coins, dice, cards, etc. For example, does the American penny have a probability of coming up heads equal to .5? Get 100 pennies and toss them from a large container 10 times to test this hypothesis.
3. Perform any one-sample binomial test of your own choosing.

EXAMPLE 10-1 Suppose we take as our population the weights of all U.S. Army lieutenants. Also, suppose it happens to be true that the mean weight μ_{pop} equals 159 pounds, and the standard deviation σ_{pop} equals 24 pounds.

PROBLEM Describe the theoretical distribution of sample means that you would get by taking many, many random samples of size 36.

SOLUTION 1. The distribution of sample means will be *normal,* since $n = 36$ is bigger than 30

2. $\mu_m = \mu_{pop} = 159$ pounds

3. $\sigma_m = \dfrac{\sigma_{pop}}{\sqrt{n}} = \dfrac{24}{\sqrt{36}} = \dfrac{24}{6} = 4$ pounds

Recall that standard deviation measures the variability in a distribution. The standard deviation of the population σ_{pop} reflects the variability among the *individual weights.* These weights range from 120 to 200. However, the standard deviation of the sample means σ_m reflects the variability among the means of samples of 36 weights and it would be quite unlikely to have a *mean* of 120 pounds in a random sample of 36 lieutenants. These means tend to vary much less and hence σ_m is smaller than σ_{pop}.
■ This distribution is represented in Fig. 10-2.

Figure 10-2

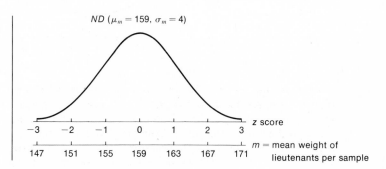

$ND\ (\mu_m = 159,\ \sigma_m = 4)$

z score

$m =$ mean weight of lieutenants per sample

ESTIMATING THE STANDARD DEVIATION OF A DISTRIBUTION OF SAMPLE MEANS

In hypothesis tests based on the theorem about sample means it is necessary to know the standard deviation of the population in order to calculate the standard deviation of the distribution of sample means. Recall that

$$\sigma_m = \frac{\sigma_{pop}}{\sqrt{n}}$$

Often, however, statisticians wish to use the theorem when they do not know σ_{pop}. They must then do two things:

1. Find an estimate for σ_{pop}. You will recall that we denote this by s.

2. Use s to get an estimate for σ_m, denoted s_m, that is,

$$s_m = \frac{s}{\sqrt{n}}$$

EXAMPLE 10-2 A claim is made that the American family, on the average, produces 5.2 pounds of organic garbage per day. A public health officer feels that the figure is probably incorrect. To test this, an experiment is set up to be analyzed at the .05 significance level. 40 families are chosen at random and their organic garbage for 1 day is weighed. The results are shown in Table 10-2.

ANALYSIS From the data the health officer computes

$$n = 40$$
$$\Sigma X = 180.3$$
$$(\Sigma X)^2 = 32,508.09$$
$$\Sigma(X^2) = 883.65$$

Step 1 H_a: $\mu_{\text{pop}} \neq 5.2$

Step 2 H_0: $\mu_{\text{pop}} = 5.2$

Step 3 $\alpha = .05$

Step 4 $n = 40 > 30$. Therefore the distribution of sample means will be approximately normal.

Step 5 Since H_a involves $\neq$, this is a two-tail test with .025 in each tail. $z_c = \pm 1.96$.
Under the assumption of the null hypothesis

$$\mu_m = \mu_{\text{pop}} = 5.2$$

$$\sigma_m = \frac{\sigma_{\text{pop}}}{\sqrt{n}}$$

Since σ_{pop} is unknown, he must compute s.

$$s = \sqrt{\frac{\Sigma(X^2) - \frac{(\Sigma X)^2}{n}}{n - 1}}$$

$$= \sqrt{\frac{883.65 - \frac{32,508.09}{40}}{40 - 1}}$$

$$= \sqrt{\frac{883.65 - 812.70}{39}}$$

$$= \sqrt{\frac{70.95}{39}}$$

$$= \sqrt{1.82}$$

Therefore, $s = 1.35$

Table 10-2	Results of Garbage-weighing Experiment for 1 Random Sample of 40 Families

family number	X, pounds of garbage
1	2.6
2	4.8
3	5.0
4	7.3
5	2.2
6	3.4
7	4.6
8	5.8
9	5.0
10	4.0
11	3.1
12	2.2
13	5.1
14	4.7
15	4.8
16	3.0
17	7.3
18	7.1
19	6.2
20	6.0
21	4.3
22	4.2
23	4.1
24	4.0
25	3.6
26	3.8
27	7.0
28	6.2
29	5.5
30	4.3
31	4.2
32	3.2
33	2.7
34	4.0
35	4.0
36	3.2
37	4.1
38	4.0
39	4.2
40	5.5
$n = 40$	$\Sigma X = 180.3$

Since σ_{pop} is unknown and is estimated by s, we estimate σ_m by s_m.

Therefore,
$$s_m = \frac{s}{\sqrt{n}} = \frac{1.35}{\sqrt{40}} = \frac{1.35}{6.32} = .21$$

This distribution is illustrated in Fig. 10-3.

Figure 10-3

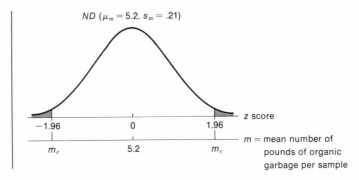

ND ($\mu_m = 5.2$, $s_m = .21$)

$$m_c = \mu_m + z_c s_m$$
$$= 5.2 + (\pm 1.96)(.21)$$
$$= 5.2 \pm .41$$
$$= 4.79 \quad \text{and} \quad 5.61$$

The decision rule is that a sample mean outside the range 4.79 to 5.61 will lead to rejection of the null hypothesis that the mean of the population is 5.2 pounds of garbage.

Step 6 The outcome of the experiment: The sample mean is

$$m = \frac{\Sigma X}{n} = \frac{180.3}{40} = 4.51 \quad \text{pounds of garbage}$$

Step 7 Conclusion The outcome 4.51 is outside the range 4.79 to 5.61. Based on this evidence, the health officer rejects the null hypothesis that the average amount of organic garbage in the population is 5.2 pounds. ■ Evidently, it is less.

EXERCISES

10-1 A commuter buys peanuts from a vending machine each evening on his way home from work. On the last 40 purchases he received the following numbers of peanuts per purchase: 12, 10, 0, 5, 15, 16, 20, 3, 12, 0, 12, 10, 9, 11, 8, 13, 15, 20, 18, 19, 20, 0, 14, 13, 15, 16, 15, 19, 11, 10, 10, 10, 3, 8, 2, 0, 0, 20, 12, 12.
(a) What is the population being sampled?
Considering these 40 purchases as his sample from the population:
(b) Compute the mean of the sample.
(c) Estimate the mean of the population.
(d) Estimate the standard deviation of the population.
(e) Describe the distribution of sample means.
(f) Estimate the mean of the distribution of sample means.
(g) Estimate the standard deviation of the distribution of sample means.
10-2 A random sample of the contributions of physicians to the United Fund was taken. 50 doctors were sampled. The results in dollars were as follows: 100, 95, 92, 92, 91, 90, 86, 85, 81, 80, 76, 76, 73, 73, 70, 70, 69, 69, 67, 66, 65, 61, 57, 52, 50, 49, 48, 47, 45, 39, 35, 35, 35, 35, 35, 30, 30, 30, 25, 25, 20, 20, 15, 15, 10, 10, 9, 5, 5, 0.
Answer Exercise 10-1 (a) to (g).

CONCLUSION

The following statements are true about the distribution of differences of sample means:

1. The distribution of differences is approximately *normal.*

2. The mean of the differences $\mu_{dm} = \mu_{\text{pop 1}} - \mu_{\text{pop 2}}$. If we are assuming for our null hypothesis that $\mu_{\text{pop 1}} = \mu_{\text{pop 2}}$, then $\mu_{dm} = 0$.

3. The standard deviation of the differences, σ_{dm}, is given by the formula

$$\sigma_{dm} = \sqrt{\frac{\sigma^2_{\text{pop 1}}}{n_1} + \frac{\sigma^2_{\text{pop 2}}}{n_2}}$$

It is usually the case that the experimenter does not know $\sigma_{\text{pop 1}}$ and $\sigma_{\text{pop 2}}$, in which case they can be estimated by s_1 and s_2. Using s_1 and s_2, we get an estimate for σ_{dm}

$$s_{dm} = \sqrt{\frac{s_1^2}{n_1} + \frac{s_2^2}{n_2}}$$

■

APPLICATION OF THE THEOREM

Let us apply the theorem to the problem of whether male scientists are paid more than female scientists during their first year of employment in industry. Recall the given information:

	male (population 1)	female (population 2)
n	100	86
m	$12,400	$11,300
s	$1,200	$1,000

We conduct a hypothesis test at the .05 significance level. We call the males population 1 and the females population 2.

Step 1 $H_a: \mu_{dm} = \mu_1 - \mu_2 > 0$. The mean salary for male scientists is larger than the mean salary for female scientists.

Step 2 $H_0: \mu_{dm} = \mu_1 - \mu_2 = 0$. The mean salaries are equal.

Step 3 $\alpha = .05$

Step 4 $n_1 = 100 > 30$ and $n_2 = 86 > 30$. Therefore, the distribution of differences is approximately normal.

Step 5 Since H_a involves $>$, this will be a one-tail test. It will be on the right because we chose the males to be population 1. So, $z_c = 1.65$.

$$\mu_{dm} = \mu_1 - \mu_2 = 0 \qquad \text{(based on } H_0\text{)}$$

$$\sigma_{dm} = \sqrt{\frac{\sigma_1^2}{n_1} + \frac{\sigma_2^2}{n_2}}$$

Since σ_1 and σ_2 are unknown, we estimate them by s_1 and s_2. Thus we estimate σ_{dm} by s_{dm} where

$$s_{dm} = \sqrt{\frac{s_1^2}{n_1} + \frac{s_2^2}{n_2}}$$

$$s_{dm} = \sqrt{\frac{(1,200)^2}{100} + \frac{(1,000)^2}{86}}$$

$$= \sqrt{\frac{1,440,000}{100} + \frac{1,000,000}{86}}$$

$$= \sqrt{14,400 + 11,627.9}$$

$$= \sqrt{26,027.9}$$

$$= 161.33$$

This distribution is illustrated in Fig. 10-4.

Figure 10-4

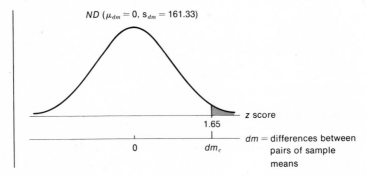

$ND\ (\mu_{dm} = 0,\ s_{dm} = 161.33)$

z score

1.65

0   dm_c

dm = differences between pairs of sample means

$$dm_c = \mu_{dm} + z_c s_{dm}$$
$$= 0 + 1.65(161.33)$$
$$= 266.19$$

The decision rule then is that if the difference between the two sample means is more than $266.19, we will say that the men are paid higher than the women.

Step 6   The experimental results are

$$m_1 - m_2 = \$12,400 - \$11,300 = \$1,100$$

Step 7   The difference between the sample means was $1,100, which is more than the critical difference, $266.19. Therefore, we claim that the average salary for the males is higher than the average salary for the females.

     Under the assumption that the two populations have the same mean, we ask: What is the probability that the difference between the mean of a randomly picked sample from population 1 and the mean of a randomly picked sample from population 2 will be $1,100? Our analysis shows that the probability of this happening is less than .05. We conclude that since we did get a difference of $1,100, then the two populations probably do not have the same mean, and the spokeswoman for women's liberation is correct.

GLOSSARY

VOCABULARY

1. Sample mean
2. Distribution of sample means
3. Sample size
4. Theorem about sample means
5. Estimate for population standard deviation
6. Difference of two sample means
7. Distribution of differences of sample means
8. Theorem about the differences of means of two samples
9. Large samples

SYMBOLS

1. μ_{pop} 2. σ_{pop} 3. s
4. m 5. μ_m 6. σ_m
7. s_m 8. dm 9. μ_{dm}
10. σ_{dm} 11. s_{dm}

FORMULAS

One-Sample Tests

1. $\mu_m = \mu_{\text{pop}}$

2. $s = \sqrt{\dfrac{\Sigma X^2 - \dfrac{(\Sigma X)^2}{n}}{n-1}}$

3. $s_m = \dfrac{s}{\sqrt{n}}$

4. $m_c = \mu_m + z_c s_m$

5. Experimental outcome, $m = \dfrac{\Sigma X}{n}$

Two-Sample Tests

6. $\mu_{dm} = \mu_1 - \mu_2$ (if H_0 states that $\mu_1 = \mu_2$, then $\mu_1 - \mu_2 = 0$)

7. $s_{dm} = \sqrt{\dfrac{s_1^2}{n_1} + \dfrac{s_2^2}{n_2}}$ 8. $dm_c = \mu_{dm} + z_c s_{dm}$

9. Experimental outcome, $dm = m_1 - m_2$

EXERCISES

10-16 A teacher used two different teaching methods in two similar statistics classes of 35 students each. Then each class took the same exam.
In one class we get $m = 82$ and $s = 4$. In the other class we have $m = 77$
and $s = 7$. Test to see if we have evidence that one method is significantly
better than the other. Use $\alpha = .05$.

10-17 In an unusual experiment, Professor Stever had some students
take an exam while hanging upside down, and another group of students
take it while lying on the floor. The results were as follows. Hanging
group: $m = 52$, $s = 10$, $n = 36$. Lying group: $m = 60$, $s = 7$, $n = 36$.
Does this indicate a significant difference in performance? Use $\alpha = .01$.

10-18 A comparison is made between two brands of toothpaste to see which does a better job of preventing cavities. In an impartial test, 31 children use Cavout and 31 use Supertooth. The results are as follows. Cavout users: Average number of new cavities = 1.6, $s = 0.7$. Supertooth users: Average number of new cavities = 2.9, $s = 0.9$. Do we have evidence that one brand is significantly better than the other? Use $\alpha = .05$.

10-19 Have the high school averages of a college's entering freshman class gone up if one year the mean high school average of 80 freshmen picked at random is 82.5 with $s = 2.5$, while the next year the mean high school average of 84 freshmen picked at random is 83.1 with $s = 2.6$? Use $\alpha = .05$.

10-20 It was once claimed by a sociologist that a woman college graduate who works is paid significantly less than her male counterpart. In testing this claim the following results were found. Women: Average salary the first year of work is $6,000 with $s = \$700$ (figures based on 32 women). Men: Average salary the first year of work is $6,850 with $s = \$600$ (figures based on 40 men). Do these figures support the claim at the .01 significance level?

10-21 It is well known that for most people the older they get, the poorer their hearing becomes. A hearing test was given to a group of 40 boys (age 10) and a group of 40 men (age 50). A high score on the test means that the person could hear high-pitched sounds. The mean score for the boys was 200 with $s = 20$. The mean score for the men was 170 with $s = 20$. Show that this is statistically significant at the .05 significance level. (Recall that the phrase "statistically significant" means that the difference is large enough to call for rejection of the null hypothesis.)

10-22 In a comparison of buying habits, the following data were obtained from two samples each consisting of 64 nuns: 10 years ago nuns bought an average of 120 habits per year with $s = 8$. Today the average is 30 habits per year with $s = 12$. Using $\alpha = .01$, does this indicate a change in the buying habits of nuns buying habits?

10-23 A school psychologist in California administered a standardized aptitude test in arithmetic to a group of 75 randomly picked sixth-grade students who had come to California from Vietnam the previous year. She gave the same test to 75 randomly picked sixth-graders who had attended California elementary schools from first grade. The mean score for the Vietnamese was 150 with $s = 25$. The California students had a mean score of 100 with $s = 40$. Show that the Vietnamese scores are significantly higher at the .01 significance level.

10-24 A group of 40 left-handed people was asked to pick up 10 pennies quickly with their right hand. Then a group of 80 right-handed people was asked to pick up 10 pennies quickly with their left hand. The length of time each person took was recorded. The following information was gathered:

$n_1 = 40$, $m_1 = 2.8$ seconds, $s_1 = 1.0$ second
$n_2 = 80$, $m_2 = 3.2$ seconds, $s_2 = 2.0$ seconds

Is this difference significant at $\alpha = .05$?

10-25 In a carefully controlled experiment, Etherea raised 35 sunflower plants by reciting a tender poem by Kahlil Gibran to each plant whenever

Table 11-3	pair number	students in first class	students in second class
	1	1	2
	2	4	3
	3	6	5
	4	8	7
	5	10	9
	6	11	12
	7	14	13
	8	16	15
	9	17	18
	10	20	19
	11	22	21
	12	23	24
	13	26	25
	14	27	28
	15	30	29
	16	31	32
	17	33	34
	18	35	36
	19	38	37
	20	39	40

The experiment has been substantially changed because each student in one class has been *paired* off with a student of similar ability in the second class. Statisticians say that this is a new *experimental design*. For this type of design it is better not to analyze these data as a two-sample t test on the difference of the sample means, because that method ignores the fact that the pairs are matched. In the case of matched pairs, the ordinary two-sample t test is a weaker approach in that it is not as likely to pick up a small but real difference between the two populations. The two-sample t test would not notice, for example, the unusual situation where every student in the first class did better than his or her "partner" in the second class, if the first class students were only a *little* better than their partners. For a somewhat less obvious illustration let us suppose that the marks for classwork were paired as shown in Table 11-4.

We would then have a sample of 20 differences. We can consider them to be a sample of size 20 from the population of all such differences. We denote these differences by the letter d. As usual, we denote the mean of a sample (of d's) by the symbol m. Assuming that the population of all d's is approximately normal, then our experiment reduces to a *one-sample* hypothesis test using sample means where our raw score is now d instead of X.

SOLUTION H_0: exemptions make no difference, $\mu_{\text{pop}} = 0$

H_a: exemptions tend to raise grades, $\mu_{\text{pop}} > 0$ (one-tail test)

$n = 20$ (therefore, we have a t test with 19 degrees of freedom)

$\mu_m = \mu_{\text{pop}} = 0$

Table 11-4	pair number	mark for student in first class (exemptions available)	mark for student in second class (no exemptions allowed)
	1	100	98
	2	96	100
	3	97	95
	4	92	90
	5	91	91
	6	93	88
	7	79	80
	8	79	83
	9	81	71
	10	86	85
	11	90	88
	12	80	77
	13	80	74
	14	82	78
	15	82	80
	16	75	71
	17	71	60
	18	73	69
	19	65	72
	20	75	65

We need to compute s_{pop} and s_m. Going back to the data above we would now have data as shown in Table 11-5.

Table 11-5	pair number	d	d^2
	1	+2	4
	2	−4	16
	3	+2	4
	4	+2	4
	5	0	0
	6	+5	25
	7	−1	1
	8	−4	16
	9	+10	100
	10	+1	1
	11	+2	4
	12	+3	9
	13	+6	36
	14	+4	16
	15	+2	4
	16	+4	16
	17	+11	121
	18	+4	16
	19	−7	49
	20	+10	100
		52	542

Our confidence interval is

$$(15.6 - 19.0) - 1.96\,(1.50) < \mu_1 - \mu_2 < (15.6 - 19.0) + 1.96\,(1.50)$$
$$-3.4 - 2.9 < \mu_1 - \mu_2 < -3.4 + 2.9$$
$$-6.3 < \mu_1 - \mu_2 < -0.5.$$

We are 95 percent sure that the average size of windows in newer houses is at least 0.5 square feet, but at most 6.3 square feet, larger than the ■ average size of windows in houses over 10 years old.

EXAMPLE 12-8 *Small Samples* Corresponding to our work in Chap. 11 on small-sample hypothesis tests, we will present an approach for the case in which the assumption that $\sigma_1 = \sigma_2$ is reasonable. You should see a more extensive text for an approach if this assumption is not reasonable.

(a) Estimate the difference between the average hat size of college presidents and the average hat size of alumni who donated over $500 to their college, if a random sample of 10 college presidents had an average hat size of 8.9 with $s = .8$, while a sample of 40 generous alumni averaged 7.1 with $s = 1.5$.

(b) Find a 95 percent confidence interval for your estimate.

SOLUTION (a) $m_1 - m_2 = 8.9 - 7.1 = 1.8$. Our best single estimate is that the average hat size of college presidents is 1.8 sizes larger than that of the alumni.

(b) $s_{dm} = \sqrt{\dfrac{s_1^2}{n_1} + \dfrac{s_2^2}{n_2}} = \sqrt{\dfrac{.64}{10} + \dfrac{2.25}{40}} = .347$

Since $n_1 = 10$ and is smaller than 30 we have a t distribution with $n_1 + n_2 - 2 = 10 + 40 - 2 = 48$ degrees of freedom. Looking up the critical t scores for the middle 95 percent of the outcomes in Table B-5, we use 50 degrees of freedom, the nearest entry at 48, and find $t_c = \pm 2.01$. Our confidence interval will be

$$1.8 - 2.01(.347) < \mu_1 - \mu_2 < 1.8 + 2.01(.347)$$
$$1.8 - .7 < \mu_1 - \mu_2 < 1.8 + .7$$
$$1.1 < \mu_1 - \mu_2 < 2.5$$

We are 95 percent sure that college presidents average at least 1.1 hat ■ sizes more than the generous alumni do.

GLOSSARY

VOCABULARY

1. Interval estimate
2. Confidence interval

FORMULAS

In a distribution of sample proportions,

1. $\mu = p$ **2.** $\sigma = \sqrt{\dfrac{pq}{n}}$

3. $\hat{\sigma} = \sqrt{\dfrac{\hat{p}\hat{q}}{n}}$ **4.** $\hat{p} - z_c\hat{\sigma} \le p \le \hat{p} + z_c\hat{\sigma}$

In a distribution of sample means,

5. $s_m = \dfrac{s}{\sqrt{n}}$

6. $m - z_c s_m \le \mu \le m + z_c s_m$ (large sample)
7. $m - t_c s_m \le \mu \le m + t_c s_m$ (small sample)

For differences between two parameters

8. $\hat{\sigma} = \sqrt{\dfrac{\hat{p}_1\hat{q}_1}{n_1} + \dfrac{\hat{p}_2\hat{q}_2}{n_2}}$

9. $(\hat{p}_1 - \hat{p}_2) - z_c\hat{\sigma} \le p_1 - p_2 \le (\hat{p}_1 - \hat{p}_2) + z_c\hat{\sigma}$
10. $(m_1 - m_2) - z_c s_{dm} \le \mu_1 - \mu_2 \le (m_1 - m_2) + z_c s_{dm}$ (large samples)
11. $(m_1 - m_2) - t_c s_{dm} \le \mu_1 - \mu_2 \le (m_1 - m_2) + t_c s_{dm}$ (small samples)

EXERCISES

12-27 A teacher desires to estimate the difference in reading levels between children from two-parent homes and children from one-parent homes in his community. Using two random samples of fourth-grade youngsters, he finds that 19 children from two-parent homes had a mean reading level m_1 of 5.1 with standard deviation $s_1 = 1.4$, and that 13 children from one-parent homes had a mean reading level m_2 of 3.8 with a standard deviation s_2 of 2.1. Find a 99 percent confidence interval for the difference in the reading levels.

12-28 The Gypsy Taxi Cab Company of Brooklyn, N.Y. is still checking tires. They now want to know if cab drivers under 25 are harder on tires than older drivers. Of their 76 cabs 32 are driven exclusively by the younger drivers and the remaining 44 by the older drivers. The younger drivers average 17,482 miles for a set of tires with a standard deviation of 1,320 miles while the older drivers average 17,728 miles with a standard deviation of 981 miles. Find a 99 percent confidence interval for the estimate of the true difference in mileage.

12-29 *Gotcha*, the consumer magazine, is testing the life of two kinds of flashlight batteries. BRITELITE claims to give more life in normal use but is more costly than ordinary batteries. They randomly purchase 50 BRITELITE batteries and 50 ordinary batteries. The BRITELITE batteries are found to have a mean life of 17.5 months of normal use with a standard deviation of 1.1 months. The ordinary batteries have a mean life of 14.7 months with a standard deviation of 1.3 months. Find a 95 percent confidence interval for the time difference in the life of the batteries.

With $\alpha = .05$ and 1 degree of freedom, $X_c^2 = 3.84$.

Decision rule: Reject the null hypothesis if $X^2 > 3.84$.
Outcome: $X^2 = 1.42$
Conclusion: We do not have enough evidence to show that sex and drinking habits of mice are related.

■

CONTINGENCY TABLES WITH ONLY ONE ROW

All the examples so far have had tables with at least two rows. We can do chi-square tests on data in contingency tables with only one row (or, equivalently, only one column). The expected values in the experiment will come from the null hypothesis and the formula for the degrees of freedom will be simpler.

EXAMPLE 13-4 A carnival wheel of fortune for a fourth-of-July fair is divided into 5 equal areas colored red, blue, red, white, and blue. The wheel is spun 50 times and the results are 25 red, 18 blue, and 7 white. Should you decide that the wheel is biased at the .05 significance level?

SOLUTION H_0: the wheel is fair
H_a: the wheel is biased

The expected values in a contingency table with only one row are *not* found by the same method that we used when we had tables of two or more rows. Instead we obtain the expected values from the null hypothesis. In this case we expect red = 2/5 (50) = 20, blue = 2/5 (50) = 20, and white = 1/5 (50) = 10.

Expected Results

red	blue	white
20	20	10

Observed Results

red	blue	white
25	18	7

We can now combine these results into a more complete table (Table 13-20) and find X^2.

Table 13-20

cell	O	E	O^2	O^2/E
1	25	20	625	31.25
2	18	20	324	16.2
3	7	10	49	4.9
	$\Sigma O = 50$	$E > 5$		$\Sigma O^2/E = 52.35$
				$\Sigma O = 50$
				$X^2 = 2.35$

■

DEGREES OF FREEDOM FOR A CONTINGENCY TABLE WITH ONE ROW

If we wish to fill in the 3 numbers in a 3×1 contingency table so that their total is 50, it should be clear that we can arbitrarily pick any 2 of them. Therefore, in our wheel of fortune problem we have 2 degrees of freedom. In general, in a $C \times 1$ contingency table there will be $(C - 1)$ degrees of freedom.

For the problem above we have

Degrees of freedom $= 3 - 1 = 2$

With $\alpha = .05$

$X_c^2 = 5.99$

Decision rule: Reject the null hypothesis if the outcome X^2 is greater than 5.99.

Outcome: $X^2 = 2.35$

Conclusion: I fail to reject the null hypothesis. The wheel may be honest.

GUIDELINES FOR A CHI-SQUARE TEST

1. Each observation in our sample falls into one and only one cell of the table.

2. The sample size is large enough so that the expected value of every cell of the table is larger than 5.

3. We compute

$$X^2 = \Sigma \left(\frac{O^2}{E} \right) - \Sigma O$$

where O is the observed result from our sample, E is the expected result, and ΣO is the sample size.

4. The experimental result X^2 as calculated above is compared to the critical value X_c^2 from Table B-7. Which value X_c^2 we choose depends on the significance level and the degrees of freedom.

5. If the experimental result X^2 is larger than our critical value X_c^2, then we reject the null hypothesis.

GLOSSARY

VOCABULARY

1. Independence of variables
2. Statistical relationship between variables
3. Observed result
4. Expected result
5. Chi square
6. Contingency table
7. Cell

14-21 [Data from Exercise 14-10(b).] Test for positive correlation.

time, seconds	distance object traveled through liquid, feet
3	9
4	16
5	25
6	36

14-22 (Data from Exercise 14-11.) Test for positive correlation.

speed, miles per hour	stopping distance, feet
30	90
40	150
50	240
60	370
70	530

14-23 Suppose you thought that there was some correlation between the length of a male college student's hair and his political beliefs. Imagine that some clever psychology professor has designed a test of political belief. When a person takes this test the score can run anywhere from 0 (extreme right-wing beliefs) to 200 (extreme left-wing beliefs). You get a random sample of 25 male students. You score each student for the two variables. These are the results.

student number	hair length, inches	test score	student number	hair length, inches	test score
1	0.5	50	14	2.5	100
2	2.0	140	15	4.5	165
3	1.0	60	16	1.5	90
4	2.5	80	17	3.0	105
5	3.0	115	18	2.5	105
6	1.5	75	19	2.0	85
7	4.5	170	20	3.5	140
8	3.5	120	21	5.0	180
9	2.5	95	22	2.5	130
10	3.0	120	23	4.0	150
11	1.0	85	24	3.0	100
12	4.0	160	25	2.0	80
13	2.0	100			

State your hypotheses. Compute r. Test for positive correlation at the .05 significance level.

14-24 An educational testing laboratory is developing a new test to measure computer programming aptitude. They wish to develop two different forms of the test. Theoretically a person should get the same score, no matter which form of the test he or she takes. To determine whether or not both forms give about the same results, they are administered to 30 people. The results were as follows:

candidate number	form A	form B	candidate number	form A	form B
1	99	80	16	67	63
2	97	95	17	67	60
3	97	87	18	65	64
4	90	88	19	65	81
5	89	83	20	65	65
6	83	90	21	63	60
7	80	85	22	62	61
8	80	78	23	61	59
9	75	40	24	60	50
10	70	76	25	59	58
11	69	70	26	50	40
12	69	71	27	43	51
13	68	70	28	40	70
14	68	72	29	20	19
15	68	68	30	3	0

Compute r. State the hypotheses. Test for positive correlation at the .01 significance level.

14-25 Benny and Hoss gamble regularly. Over the past year their profits in hundreds of dollars have been as follows:

month	Benny	Hoss	month	Benny	Hoss
January	1	2	July	−4	−1
February	3	−5	August	1	4
March	10	0	September	0	7
April	−5	1	October	3	2
May	2	3	November	8	1
June	1	8	December	2	3

(a) Test at the .05 significance level to see if there is a difference between the average amount Benny wins per month and the average amount Hoss wins per month.

(b) Test at the .05 significance level to see if there is any correlation between Benny's monthly profit and Hoss's monthly profit.

14-26 The following data were collected at random from Ms. Betty's School for Young People.

student	reading	spelling	math	music
Sam	20	7	100	10
Samantha	15	7	70	. . .
Toni	25	10	60	3
Anthony	35	8	90	9
Salvatore	30	9	. . .	20
Sally	50	8	80	15
Pat	40	10	80	5

Find the coefficient of correlation, and perform a hypothesis test at the .05 significance level for the following:

Tests Involving Variance

15   The hypotheses we have tested so far were either about means or proportions and these have been quite useful. There are also many other important statistical hypotheses which are about variability. Two common measures of variability used in hypothesis tests are the *standard deviation* and the square of the standard deviation, the *variance.* You have already worked quite a bit with the standard deviation. One of its advantages is that it measures variability in whatever units the variable is measured in. For example, we might say that the average distance from the floor to the doorknobs in a certain apartment house is 36.5 *inches,* with a standard deviation of .25 *inch.* In contrast to this if you use the variance to measure variability, the variance is not in the original units of the problem and so is not as easy to interpret intuitively. In the example just stated, the variance is (1/4)², or 1/16, but the unit is "inches squared" which makes no intuitive sense. In this chapter, however, we will see that certain hypotheses should be tested by looking at sample *variances* rather than sample standard deviations, because statistics based on the sample variances can be meaningfully compared with certain well-established tables of critical values. We do not have equivalent tables for statistics based on the sample standard deviations. Note that where we use the letter s to represent standard deviations calculated from samples, we will use s^2 to represent variances calculated from samples. Our basic formulas are

$$s = \sqrt{\frac{\Sigma X^2 - \frac{(\Sigma X)^2}{n}}{n - 1}} \quad \text{and} \quad s^2 = \frac{\Sigma X^2 - \frac{(\Sigma X)^2}{n}}{n - 1}$$

In this chapter we are going to illustrate three types of problems. The first is a test of a claim about the variance of a population. For example, a claim might say that the heights in inches of 6-year-old children in the Los Angeles public school system have a variance equal to 4. We would have H_0: variance = 4. In symbols, this is H_0: $\sigma^2 = 4$. This is equivalent to claiming that the standard deviation is 2 inches. This type of claim will be tested by computing a statistic and comparing it with a critical value from a *chi-square* table. (So this is a new application of the chi-square table.) We will call these tests *one-sample tests of variance.*

The second type of problem will be a test to compare the variances of *two* populations. An example might be to test the claim that in sea-farming lobsters one diet produces more *erratic* size lobsters than another diet. The null hypothesis will be H_0: $\sigma_1^2 = \sigma_2^2$. The computations will involve computing the variances of *two* samples, one from each population, then using these to compute a statistic which can be compared to critical values in a new table called the F table. We will call these tests *two-sample comparisons of variance.*

The third type of problem will use the F table and sample variances to analyze the hypothesis that the *means* of several populations are equal. This type of test is an extension of the two-sample mean tests already developed in Chap. 11. We will call these tests *comparisons of the means of several populations.*

ONE-SAMPLE TESTS OF VARIANCE

The Deep Dark Device Department of the Kynda Klever Kamera Company manufactures darkroom timers—"the kind you wind." Sample Sam, the quality control man, always tests every timer they make to see if the assembly process is working properly. Sam does the test by setting each timer to the "30 seconds" setting, and then timing it electronically to see how long it actually takes to ring. He repeats this 11 times for each timer. He knows from past experience that if he would test any particular timer that is working properly *thousands* of times and make a frequency chart and histogram of his results, then he would see an approximately normal distribution of times with a mean of 30 seconds (Fig. 15-1).

Figure 15-1

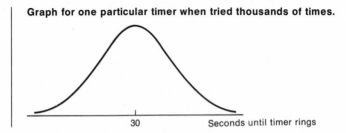

Graph for one particular timer when tried thousands of times.

30   Seconds until timer rings

Sam's concern is: Does this distribution have a comparatively small variance? This means that this particular timer behaves consistently and is therefore more reliable in the darkroom. For example, if for two particular timers his tests resulted in the two graphs shown in Fig. 15-2, this would show that timer A is more reliable than timer B.

Let $p = P$(a plus sign)

H_0: the number of plus signs is the same as the number of minus signs, $p = .5$

H_a: there will be more plus than minus signs, $p > .5$ (one-tail test)

$\mu = np = 11(.5) = 5.5$

$\sigma = \sqrt{npq} = \sqrt{11(.5)(.5)} = 1.66$

Since $np = nq = 5.5 > 5$ we have a normal distribution of successes and for $\alpha = .01$ the critical value is $z_c = +2.33$.   Therefore,

$S_c = \mu + z\sigma = 5.5 + 2.33(1.66) = 8.4$

Our decision rule will be to reject H_0 if we get more than 8.4 plus signs. Since our outcome is 6 plus signs, we fail to reject H_0.   We have failed to show that reading skills have decreased after 1 month. ■

EXAMPLE 16-2 A class was given a test under careful supervision.   Although cheating was not specifically mentioned, it was almost certain that nobody could or did cheat.   The next day the class was *told* that because some of them cheated, the test would be given again with several proctors in the room. The marks for both tests were as shown in Table 16-3.

Table 16-3

student	X, 1st day	Y, 2nd day	student	X, 1st day	Y, 2nd day
Ellington	70	66	Mancini	84	84
Lombardo	43	39	Alpert	78	70
Prima	91	85	Severinsen	92	69
Dorsey	89	92	Welk	83	84
Basie	73	72	Floren	75	74
James	64	63	Cugat	73	72
Kostelanetz	51	40	Duchin	89	83
Bach	83	88			

Did the announcement lower the test results?   Use $\alpha = .05$.

SOLUTION Let $p = P(X > Y)$

H_0: the announcement did not affect the grades, $p = .5$

H_a: the announcement lowered the grades, $p > .5$ (one-tail test)

Our data can now be shown as in Table 16-4.

Table 16-4

student	sign	student	sign
Ellington	+	Mancini	0
Lombardo	+	Alpert	+
Prima	+	Severinsen	+
Dorsey	−	Welk	−
Basie	+	Floren	+
James	+	Cugat	+
Kostelanetz	+	Duchin	+
Bach	−		

Ignoring the pair of scores for Mancini, which were the same, we have $n = 14$, $np = nq = 7 > 5$, $\mu = 7$, $\sigma = \sqrt{7(.5)} = 1.87$. Therefore, $S_c = 7 + 1.65(1.87) = 10.1$. Our decision rule will be to reject the null hypothesis if our outcome is more than 10.1 plus signs. Our outcome is 11 plus signs and we can reject the null hypothesis; the evidence indicates that the announcement lowered the test grades.

EXERCISES

16-1   All freshmen are weighed by the medical office when school opens in September.   Prof. Haman Deggs teaches a class in nutrition.   At the end of the semester in December he asked his students to be weighed again.   Testing at the .05 significance level, are the variations in weight shown random?

(a) Test by the two-sample sign test.

(b) Assume that weights are distributed normally and test by a matched-pairs t test.

student number	September weight	December weight	student number	September weight	December weight
1	140	137	11	75	81
2	112	110	12	210	212
3	176	210	13	193	189
4	98	98	14	145	140
5	180	165	15	144	142
6	140	145	16	139	139
7	154	150	17	180	164
8	193	185	18	165	159
9	128	129	19	98	97
10	102	101	20	141	136

16-2   Do premed students at Leach College score better in physics than in math?   The final exam scores for a random sample of these students are given below.   Use $\alpha = .05$.

(a) Test using the two-sample sign test.

(b) Assume the grades are normally distributed and use the matched-pairs t test.

student	math	physics	student	math	physics
Korn, A.	90	95	Lash, I.	78	78
Tropey, N.	78	76	Ective, F.	90	60
Shorr, C.	80	83	Sera, K.	87	94
Frost, D.	81	82	Bow, L.	78	83
Lope, E.	94	90	Knott, Y.	72	48
Kupp, T.	30	31	Cleaf, O.	99	98
Nee, G.	63	60	Kann, P.	70	80
Bohr, R.	70	78	Kneeaform, Q.	62	78

16-3   Two efficiency experts independently rated the performance of a group of assembly-line workers.   Their ratings were as follows.

worker	first expert rating	second expert rating	worker	first expert rating	second expert rating
Alice	8	3	Jorja	10	5
Pat	5	9	Jean	8	7
Sam	2	8	Jo	7	8
Dee	6	7	Carly	6	6
Tony	7	5	Mary	9	2
Sal	8	4	Bill	3	10
Fran	9	4	Dave	4	9

Using $\alpha = .01$ can we conclude that the two experts use the same standards in rating workers?

16-4 70 of Mr. Abel Riemann's pupils in geometry were in his algebra classes last year. To compare their grades in these two different subjects he converted their grades to z scores and then found the differences of their algebra grades minus their geometry grades. He got 40 plus signs, 25 minus signs, and 5 zeros. Test at the .05 significance level.

16-5 Given that 13 Plutans have more arms than legs, 32 have more legs than arms, and 3 had the same number of arms and legs, test at $\alpha = .01$ whether the mean number of arms is different from the mean number of legs.

16-6 Since half of the scores fall above the median and half below, we can use reasoning similar to that used in a two-sample sign test to test the hypothesis that the median of a population is some given number. A plus sign corresponds to a score above the median and a minus sign corresponds to a score below the median. Use this approach to solve the following problem.

Eve Wormwood packages apples with a median of 8.5 apples per box. Wormwood samples some packages of her competitor, Adam Upright. She finds the following numbers of apples per box: 10, 12, 8, 7, 15, 9, 8, 7, 12, 8, 8, 9, 9, 9, 9, 10, 12, 7, 8, 15, 14, 18, 12, 6, 8, 9, 9, and 8. Should Wormwood be convinced that Upright's median is greater than hers? Use $\alpha = .05$.

16-7 Is the median age of teachers at Gray University 62 years? Test at $\alpha = .05$ if a sample of teachers had the following ages: 25, 47, 53, 53, 58, 59, 61, 62, 62, 65, 66, 66, 66, 73, 81, 85, and 94.

16-8 Lloyd is the attendant on a drawbridge from 10 P.M. to 6 A.M. His job is not exciting. He notices that when he raises the bridge it seldom interferes with any traffic, even though he may raise the bridge several times per night. He wonders if the median of the number of times in a *week* that any traffic stops for the raised bridge is less than 1. In the past 60 weeks he noticed that in 30 of the weeks no traffic was stopped, in 20 of the weeks traffic was stopped once, and in the remaining 10 weeks traffic was stopped more than once. Test at the .01 significance level.

16-9 Given a paired-sign test with 20 pairs. In a one-tail test on the right, what is the smallest number of plus signs that will cause a rejection of H_0 if $\alpha = .05$?

THE RUNS TEST

Ida Noh, a safety expert, has been monitoring her radar set behind a billboard. Each time a car passes doing the speed limit or less she writes S for slow. Each time a car passes doing more than the speed limit she writes F for fast. The results after 40 cars were

S S F F F F S S S S S S S S S S F S S F F F F S
S S S S S S F F F F S S S S S F S S.

Ida wants to know whether or not speeders and nonspeeders occur randomly. That is, do speeders tend to come bunched together? She breaks the series of outcomes into **runs** of S's and **runs** of F's as follows:

S S   F F F F   S S S S S S S S   F   S S   F F F F
S S S S S S   F F F   S S S S   F   S S

She has 11 runs. If we let n_1 equal the number of S's, n_2 equal the number of F's, and U equal the number of runs, then $n_1 = 26, n_2 = 14$, and $U = 11$.

This is just one possible way that her string of 26 S's and 14 F's might have turned out. Some other arrangement would result, perhaps, in a different number of runs. So we can talk about U, the number of runs, as a random variable. If we repeatedly got random arrangements of 26 S's and 14 F's we would end up with a *distribution of U's*.

It can be shown that the *mean* number of runs, when you *randomly* arrange n_1 items of one kind and n_2 items of another kind, is

$$\mu_U = \frac{2n_1n_2}{n_1 + n_2} + 1$$

In this case

$$\mu_U = \frac{2(26)(14)}{26 + 14} + 1 = 19.2$$

The standard deviation of the number of runs is

$$\sigma_U = \sqrt{\frac{2n_1n_2(2n_1n_2 - n_1 - n_2)}{(n_1 + n_2)^2(n_1 + n_2 - 1)}}$$

In this case

$$\sigma_U = \sqrt{\frac{2(26)(14)[2(26)(14) - 26 - 14]}{(26 + 14)^2(26 + 14 - 1)}}$$

$$= \sqrt{8.03} = 2.83$$

Further, the distribution of U's is approximately normal if both n_1 and n_2 are greater than 10. Therefore, Ms. Noh has

$ND(\mu_U = 19.2, \sigma_U = 2.83)$

H_0: fast and slow cars arrive randomly

H_a: fast and slow cars do not arrive randomly

This is a two-tail test since there could be either too many runs or too few runs if they are not random. Testing at the .05 significance level, we have

STATISTICS

TEST A - PART II

1. The secretary of an association of professional landscape gardeners claims that the average cost of services to customers if $90 per month. Feeling that this figure is too low, we question a random sample of 10 customers. Our sample yields a mean cost of $125 with s = $20. Test at the .05 significance level. Assume that such costs are normally distributed.

2. Data were collected on 10 adults who enrolled in a weight-losing program. Their weights were recorded before and after the program. For the 10 pairs of differences m was 5.8 pounds lost with s = 5. Using α = .05 is this evidence that the program works?

3. Because of a great number of applicants the director of admissions at a private university

Table B-9

Critical Values of r for a Two-Tail Test

(Values of r are given in this table without signs. All values are both positive and negative, that is, $r_c = \pm 1.00$.)

n	r_c for $\alpha = .05$	r_c for $\alpha = .01$
3	1.00	1.00
4	.95	.99
5	.88	.96
6	.81	.92
7	.75	.87
8	.71	.83
9	.67	.80
10	.63	.76
11	.60	.73
12	.58	.71
13	.53	.68
14	.53	.66
15	.51	.64
16	.50	.61
17	.48	.61
18	.47	.59
19	.46	.58
20	.44	.56
21	.43	.55
22	.42	.54
23	.41	.53
24	.40	.52
25	.40	.51
26	.39	.50
27	.38	.49
28	.37	.48
29	.37	.47
30	.36	.46

For values of r_c, when n is greater than 30 use

$$r_c = \frac{t_c}{\sqrt{t_c^2 + (n - 2)}}$$

where t_c is the corresponding critical value of t for $(n - 2)$ degrees of freedom in Table B-5.

Table B-10

Critical Values of r for a One-Tail Test

(Values of r are given in this table without signs. You must determine whether the critical values of r are positive or negative from the alternative hypothesis.)

n	r_c for $\alpha = .05$	r_c for $\alpha = .01$
3	.99	1.00
4	.90	.98
5	.81	.93
6	.73	.88
7	.67	.83
8	.62	.79
9	.58	.75
10	.54	.72
11	.52	.69
12	.50	.66
13	.48	.63
14	.46	.61
15	.44	.59
16	.42	.57
17	.41	.56
18	.40	.54
19	.39	.53
20	.38	.52
21	.37	.50
22	.36	.49
23	.35	.48
24	.34	.47
25	.34	.46
26	.33	.45
27	.32	.45
28	.32	.44
29	.31	.43
30	.31	.42

For values of r_c, when n is greater than 30 use

$$r_c = \frac{t_c}{\sqrt{t_c^2 + (n - 2)}}$$

where t_c is the corresponding critical value of t for $(n - 2)$ degrees of freedom in Table B-6.

Critical Values of F
for $\alpha = .05$
for a One-Tail Test

Degrees of freedom for numerator

	1	2	3	4	5	6	7	8	9	10	12	15	20	24	30	40	50	∞
1	161	200	216	225	230	234	237	239	241	242	244	246	248	249	250	251	252	254
2	18.5	19.0	19.2	19.2	19.3	19.3	19.4	19.4	19.4	19.4	19.4	19.4	19.4	19.5	19.5	19.5	19.5	19.5
3	10.1	9.55	9.28	9.12	9.01	8.94	8.89	8.85	8.81	8.79	8.74	8.70	8.66	8.64	8.62	8.59	8.58	8.53
4	7.71	6.94	6.59	6.39	6.26	6.16	6.09	6.04	6.00	5.96	5.91	5.86	5.80	5.77	5.75	5.72	5.70	5.63
5	6.61	5.79	5.41	5.19	5.05	4.95	4.88	4.82	4.77	4.74	4.68	4.62	4.56	4.53	4.50	4.46	4.44	4.37
6	5.99	5.14	4.76	4.53	4.39	4.28	4.21	4.15	4.10	4.06	4.00	3.94	3.87	3.84	3.81	3.77	3.75	3.67
7	5.59	4.74	4.35	4.12	3.97	3.87	3.79	3.73	3.68	3.64	3.57	3.51	3.44	3.41	3.38	3.34	3.32	3.23
8	5.32	4.46	4.07	3.84	3.69	3.58	3.50	3.44	3.39	3.35	3.28	3.22	3.15	3.12	3.08	3.04	3.03	2.93
9	5.12	4.26	3.86	3.63	3.48	3.37	3.29	3.23	3.18	3.14	3.07	3.01	2.94	2.90	2.86	2.83	2.80	2.71
10	4.96	4.10	3.71	3.48	3.33	3.22	3.14	3.07	3.02	2.98	2.91	2.85	2.77	2.74	2.70	2.66	2.64	2.54
11	4.84	3.98	3.59	3.36	3.20	3.09	3.01	2.95	2.90	2.85	2.79	2.72	2.65	2.61	2.57	2.53	2.50	2.40
12	4.75	3.89	3.49	3.26	3.11	3.00	2.91	2.85	2.80	2.75	2.69	2.62	2.54	2.51	2.47	2.43	2.40	2.30
13	4.67	3.81	3.41	3.18	3.03	2.92	2.83	2.77	2.71	2.67	2.60	2.53	2.46	2.42	2.38	2.34	2.32	2.21
14	4.60	3.74	3.34	3.11	2.96	2.85	2.76	2.70	2.65	2.60	2.53	2.46	2.39	2.35	2.31	2.27	2.24	2.13
15	4.54	3.68	3.29	3.06	2.90	2.79	2.71	2.64	2.59	2.54	2.48	2.40	2.33	2.29	2.25	2.20	2.18	2.07
16	4.49	3.63	3.24	3.01	2.85	2.74	2.66	2.59	2.54	2.49	2.42	2.35	2.28	2.24	2.19	2.15	2.13	2.01
17	4.45	3.59	3.20	2.96	2.81	2.70	2.61	2.55	2.49	2.45	2.38	2.31	2.23	2.19	2.15	2.10	2.08	1.96
18	4.41	3.55	3.16	2.93	2.77	2.66	2.58	2.51	2.46	2.41	2.34	2.27	2.19	2.15	2.11	2.06	2.04	1.92
19	4.38	3.52	3.13	2.90	2.74	2.63	2.54	2.48	2.42	2.38	2.31	2.23	2.16	2.11	2.07	2.03	2.00	1.88
20	4.35	3.49	3.10	2.87	2.71	2.60	2.51	2.45	2.39	2.35	2.28	2.20	2.12	2.08	2.04	1.99	1.96	1.84
25	4.24	3.39	2.99	2.76	2.60	2.49	2.40	2.34	2.28	2.24	2.16	2.09	2.01	1.96	1.92	1.87	1.84	1.71
30	4.17	3.32	2.92	2.69	2.53	2.42	2.33	2.27	2.21	2.16	2.09	2.01	1.93	1.89	1.84	1.79	1.76	1.62
40	4.08	3.23	2.84	2.61	2.45	2.34	2.25	2.18	2.12	2.08	2.00	1.92	1.84	1.79	1.74	1.69	1.66	1.51
50	4.03	3.18	2.79	2.56	2.40	2.29	2.20	2.13	2.07	2.02	1.95	1.87	1.78	1.74	1.69	1.63	1.60	1.44
∞	3.84	3.00	2.60	2.37	2.21	2.10	2.01	1.94	1.88	1.83	1.75	1.67	1.57	1.52	1.46	1.39	1.35	1.00

Degrees of freedom for denominator

Table B-15 (*Continued*)

Row number	Column number				
	6	7	8	9	10
1	59585	55855	82365	01993	32159
2	99315	47346	46747	13961	62746
3	71946	66550	71398	09429	78822
4	55933	33377	15145	53129	21927
5	91302	68175	42524	84064	99081
6	69301	57163	74375	73120	31804
7	69703	37390	92324	81465	15299
8	30142	98101	56566	91609	22162
9	94357	07887	58787	95481	47497
10	98711	39491	44629	86428	04817
11	18876	76639	02557	14488	44599
12	11041	49789	28505	26817	79868
13	44531	60149	38185	27649	65689
14	35201	13859	11417	54669	91942
15	51496	00702	15377	81154	11253
16	77536	14221	66118	15785	36450
17	80105	22615	50248	84476	04437
18	38927	62300	41144	41122	15834
19	62095	38486	06567	66619	22291
20	21033	79678	17061	78044	45816
21	01861	05443	27009	82367	77023
22	82546	29344	12717	71569	11650
23	07704	31397	14407	68538	47245
24	45438	96942	41490	95625	74372
25	73080	31288	62998	50229	59228
26	51213	22022	17542	34292	27046
27	07853	33341	39678	97069	18147
28	67157	79292	12062	13059	01023
29	19543	35308	15248	29468	89334
30	42586	09867	59525	80071	22179
31	44575	10724	45663	74867	29527
32	15097	02514	13933	37379	17175
33	29531	65157	03288	73990	68121
34	66820	24909	55072	22189	69704
35	62400	42453	58142	37090	38416
36	30664	04885	44759	38114	26737
37	53826	80992	09150	50202	58871
38	46572	61680	33090	36425	54778
39	68360	19934	65396	31392	64347
40	67764	87188	14696	47297	46112
41	05709	55470	77366	37009	87365
42	16997	77210	34986	61062	75057
43	31986	47074	18206	17931	89350
44	92804	12704	46401	59458	04209
45	60961	73747	14967	25818	41885
46	25752	91032	14664	96881	40697
47	85310	40277	29855	69362	57957
48	84690	07214	75029	31629	42432
49	57864	33479	91483	20535	73211
50	32944	85176	38532	07162	74355

Answers to Selected Odd-numbered Exercises†

CHAPTER 1

1-3(a) Statistical inference; (b) statistical inference; (c) descriptive statistics; (d) statistical inference; (e) descriptive statistics

CHAPTER 2

2-1(a) 35; (b) $4 + 9 + 16 + 25 + 36 + 49 + 64 = 203$; (c) $(35)^2 = 1,225$; (d) 5; (e) 21; (f) 0; (g) 4.67; (h) 4.67

2-3(a) 12; (b) 2; (c) 0; (d) 144; (e) 46; (f) 4.4

2-5(a) 14.96; (b) 14.96   **2-7** 55; 68; 72, 70, and 68

2-9 3.3; 3.25; no mode   **2-11** Unlimited

2-13 85.56   **2-15** 90.56   **2-17** 269.44

2-19(a) 2.89; (b) 6; (c) $\overline{X} = \dfrac{\Sigma X}{n}$

2-21(a) There are none; (b) 3, 20

2-23(a) 6.8, 11.2, 3.35; (b) 68, 1, 120, 33.5; (d) there are none

2-25 $s^2 = 0.153$; $s = 0.39$ **2-27** No

2-31 15.75, 14.72; (a) 7; (b) 58%; (c) 12; (d) 100%

2-33 13,521.17 pesos; 425.30 pesos

2-35 $\mu_X = -\mu_Y$; $\sigma_X = \sigma_Y$

2-39 Pythagorean

2-43 False

---

† Answers may sometimes vary slightly depending on the degree of accuracy and the number of decimal places carried through your solution. Do not be concerned over minor differences between your calculations and those given here.

16-7 $S_c = 15(.5) \pm 1.96 \sqrt{15(.5)(.5)} = 3.9$ and 11.3; $S = 8$ plus signs. Fail to show that the median is not equal to 62.

16-9 $S_c = 20(.5) + 1.65\sqrt{20(.5)(.5)} = 13.7$. The smallest number of plus signs is 14.

16-11(a) $n_1 = 7$, $n_2 = 4$, $U = 7$, $\mu_U = 2(7)(4)/7 + 4 + 1 = 6.1$, $\sigma_U = \sqrt{\dfrac{2(7)(4)[2(7)(4) - 7 - 4]}{(7 + 4)^2(7 + 4 - 1)}} = 1.4$; (c) $n_1 = 9$, $n_2 = 12$, $U = 6$, $\mu_U = 2(9)(12)/9 + 12 + 1 = 11.3$, $\sigma_U = \sqrt{\dfrac{2(9)(12)[2(9)(12) - 9 - 12]}{(9 + 12)^2(9 + 12 - 1)}} = 2.2$;

(e) $n_1 = 11$, $n_2 = 8$, $U = 13$; $\mu_U = 2(11)(8)/11 + 8 + 1 = 10.3$; $\sigma_U = \sqrt{\dfrac{(2)(11)(8)[2(11)(8) - 11 - 8]}{(11 + 8)^2(11 + 8 - 1)}} = 2.1$

16-13 $U_c = \left[\dfrac{2(11)(11)}{11 + 11} + 1\right] \pm 1.96$

$\sqrt{\dfrac{2(11)(11)[2(11)(11) - 11 - 11]}{(11 + 11)^2(11 + 11 - 1)}} = 7.5$ and 16.5; $U = 8$. Fail to show that variations from the median are not random.

16-15(a)

	U
TFFFF	2
FTFFF	3
FFTFF	3
FFFTF	3
FFFFT	2

(b) $\mu = 13/5 = 2.6$, $\mu = 2(1)(4)/1 + 4 + 1 = 2.6$;

(c) $\sigma = \sqrt{\dfrac{35 - 13^2/5}{5}} = \sqrt{.24} = .490$,

$\sigma = \sqrt{\dfrac{2(1)(4)[2(1)(4) - 1 - 4]}{(1 + 4)^2(1 + 4 - 1)}} = \sqrt{.24} = .490$

ANSWERS TO EXERCISES IN APPENDIX A

A-1(a) 23.73, (b) 4.97, (c) .046, (d) 1.202, (e) 9.628, (f) .0012, (g) 21.25, (h) .5, (i) .02, (j) 0

A-2(a) -2.33; (b) .081, .41, .6, 4.51, 4.7; (c) $-.273$, .273, .41

A-3(a) True, (b) true; (c) false; (d) true; (e) 5, 6; (f) 4, 5, 6; (g) 0, 1, 2, 3, 4; (h) 0, 1, 2, 3, 4, 5; (i) 3, 4; (j) 2, 3, 4, 5

A-4(a) 5%; (b) .3%; (c) .37; (d) .032; (e) .375, 37.5%; (f) .263, 26.3%; (g) 11.5; (h) 8; (i) 30%; (j) 25%

A-5(a) -15, (b) -15.35, (c) -16.5, (d) -2, (e) -5

A-6(a) 7.446, (b) -41, (c) 11.167, (d) 6.92 and $-.92$, (e) true

A-7(a) $+16$, (b) .343, (c) 1/16, (d) .3087

A-8(a) 2, (b) 6.325, (c) 1.712, (d) 5.413, (e) .5413, (f) 19.34, (g) .1934, (h) .06116

A-9(a) 1.550, (b) 2.315, (c) 4.45

Index

3. $s_m = \dfrac{s}{\sqrt{n}}$

4. $m_c = \mu_m + z_c s_m$

5. Experimental outcome, $m = \dfrac{\Sigma X}{n}$

TWO-SAMPLE TESTS

6. $\mu_{dm} = \mu_1 - \mu_2$ (if H_0 states that $\mu_1 = \mu_2$, then $\mu_1 - \mu_2 = 0$)

7. $s_{dm} = \sqrt{\dfrac{s_1^{\,2}}{n_1} + \dfrac{s_2^{\,2}}{n_2}}$

8. $dm_c = \mu_{dm} + z_c s_{dm}$

9. Experimental outcome, $dm = m_1 - m_2$

Chapter 11

ONE-SAMPLE TESTS

1. Degrees of freedom $= n - 1$

2. $\mu_m = \mu_{\text{pop}}$

3. $s_m = \dfrac{s}{\sqrt{n}}$

4. $m_c = \mu_m + t_c s_m$

5. Experimental outcome, $m = \dfrac{\Sigma X}{n}$

TWO-SAMPLE TESTS

1. Degrees of freedom $= n_1 + n_2 - 2$

2. $\mu_{dm} = \mu_1 - \mu_2$ (if H_0 states that $\mu_1 = \mu_2$, then $\mu_1 - \mu_2 = 0$)

3. $s_{dm} = \sqrt{\dfrac{s_1^2}{n_1} + \dfrac{s_2^2}{n_2}}$

4. $dm_c = \mu_{dm} + t_c s_{dm}$

5. Experimental outcome, $dm = m_1 - m_2$

Chapter 12

IN A DISTRIBUTION OF SAMPLE PROPORTIONS

1. $\mu = p$

2. $\sigma = \sqrt{\dfrac{pq}{n}}$

3. $\hat{\sigma} = \sqrt{\dfrac{\hat{p}\hat{q}}{n}}$

4. $\hat{p} - z_c \hat{\sigma} \le p \le \hat{p} + z_c \hat{\sigma}$

IN A DISTRIBUTION OF SAMPLE MEANS

5. $s_m = \dfrac{s}{\sqrt{n}}$